the
1000
most-asked
gardening
questions

An Hachette UK Company
www.hachette.co.uk

First published in Great Britain in 2009 by Spruce,
an imprint of Octopus Publishing Group Ltd
Carmelite House, 50 Victoria Embankment, London EC4Y 0DZ
www.octopusbooks.co.uk

This edition published in 2022 by Pyramid, an imprint of Octopus Publishing Group Ltd.

ISBN 978-0-7537-3503-9

A CIP catalogue record for this book is available from the British Library

Printed and bound in China

10 9 8 7 6 5 4 3 2 1

Publisher: Lucy Pessell
Designer: Hannah Coughlin
Junior Editor: Sarah Kennedy
Editorial Assistant: Emily Martin
Production Manager: Caroline Alberti

This book contains the opinions and ideas of the author. It is intended to provide helpful
and informative material on the subjects addressed in this book and is sold with the under-
standing that the author and publisher are not engaged in rendering any kind of personal
professional services in this book. The author and publisher disclaim all responsibility for
any liability, loss or risk, personal or otherwise, which is incurred as a consequence, directly
or indirectly, of the use and application of any of the contents of this book.

the
1000
most-asked
gardening
questions

Daphne Ledward

Contents

Introduction

Look in any gardening book or at the Q & A pages of any gardening magazine over the last 50 years or so, and you may be surprised to see that most of the questions they contain are ones we're still asking today. Over the decades, the same subjects come up time and time again. Where there is a difference, however, is in how these problems are tackled by today's gardener.

The gardener of a few decades ago generally approached the subject in a different way from the garden owner of today. Gardening was a dedicated hobby, more time was available for tending the plot, there were fewer, other more attractive interests, and space for cultivation was not at a premium. The modern gardener, on the other hand, has to juggle the care of an often small urban garden with home improvements, a tempting choice of other evening and weekend activities, and a barrage of television makeover features that rely heavily on fashion and style while offering little practical advice.

In addition, the environmental issue plays an increasingly large part in how gardening queries are tackled. When I started working on gardening question programmes at the beginning of the 1980s, the questioner would usually want a solution that was quick and effective, regardless of how this may have affected the natural world. In the ten years after the banning of the pesticides DDT and Aldrin had made the general public aware of the long-term damage to the ecosystem caused by the use of such persistent and toxic chemicals, it was still easy for the amateur gardener to access a multitude of readily available chemical potions that were likely to have persistent harmful effects.

Thankfully, over the following 30 years, the general public awareness of and attitude toward the environment improved, and we saw the withdrawal of many chemicals that had a seriously detrimental effect on wildlife and long-term health. This has led to an enormous change in attitude for the modern gardener. He or she still wants a solution to a particular problem, but now this must be the 'greenest' one available.

Most modern gardeners are concerned about the carbon footprint of imported fruit and vegetables, and the heavy reliance on agrochemicals to maximize cropping. While this move toward homegrown food appears to be a return to a way of life that prevailed before instant supermarket availability, today many gardeners who desire self-sufficiency have minimal space for food cultivation, so techniques may, out of necessity, be very different.

Our 1000 most-asked gardening questions may remain largely the same as those asked in our grandparents' day, but the answers can be very different. Time, space, cost and global impact are of the essence; the 21st-century gardener wants a simple, inexpensive, ergonomic, understandable and 'green' answer to a straightforward question – and that is exactly what this book hopes to achieve.

Good luck, and good digging.

spring

"When is the best time to dig vacant land?"

Dig it in autumn so that the winter weather can break it down.

"I am making a new garden on old pasture and the ground is full of grubs. What should I do?"

Dig over the areas in which you want to grow vegetables and perennials and leave them rough for a week or two. The birds will do the rest.

"Is a stainless steel spade essential?"

No it isn't, but if you have one, you must remember to clean and oil it after each use. In general, spend as much as you can afford on your garden tools.

"What is a potato fork?"

A potato fork has flat tines (prongs) that do not damage potatoes when you are lifting them. This type of fork is a useful alternative to a spade for digging heavy soil.

"When is a rotavator useful?"

A rotavator will turn over clean soil and break it down into evenly sized pieces, but you should only use one on ground that you have already cleared of perennial weeds, otherwise you will leave pieces of their roots in the ground, which will reshoot.

"What is the minimum depth I should dig my vegetable garden?"

A depth of 25–30 cm (10–12 in) is plenty for most crops.

"What is meant by topsoil?"

This is the fertile top layer of your soil. It contains humus (rotted organic material) and beneficial organisms.

"What is meant by 'no-dig' cultivation?"

This is when you spread a thick layer of garden compost or farmyard manure on the surface of the ground and leave it for worms and other soil organisms to work in. It gives the soil a light and easy-to-work texture.

"What is subsoil?"

This is the infertile layer of soil underlying the topsoil. You can break it up when you are digging to improve drainage, but take care not to bring it to the top.

"Why should I put lime on the soil?"

Lime "sweetens" a sour soil, helping to make nutrients available, and increases the alkalinity. Many plants, such as dianthus, clematis and cabbages, grow best in a limey soil.

"How can I help my sandy soil retain moisture?"

Add plenty of manure or compost every year.

"Is it true that I should not put manure and lime on the soil at the same time?"

Lime releases ammonia gas from the manure, making it less effective as a plant food. The usual advice is to add lime in autumn and manure in spring.

"How can I improve a heavy clay soil?"

Keep adding compost or manure, sharp sand or horticultural grit, and the texture will start to improve.

"Can I use garden soil in containers?"

Use a proprietary compost. Garden soil might contain pests and diseases.

"I find digging physically difficult. How can I rid the soil of weeds before planting without using chemicals?"

Cover the area with black polythene and most weeds will disappear completely in 12 months.

"Are worms beneficial to the soil?"

Yes. They work in, and break down, organic material, improving the soil's quality.

"I want to make a lawn on stony soil, but the stones keep coming to the top."

Sow the grass, then, before the first cut, pick off as many stones as you can. Once the grass is established, the stones will stay underneath.

"What is a mulch?"

A mulch is a soil covering that suppresses weeds and reduces evaporation. Good examples are chipped bark, gravel and black polythene.

"Which is better, light or dark soil?"

A dark soil warms up more quickly in spring. The more organic material a light soil receives, the darker it will eventually become.

"Is ash good for the soil?"

Bonfire and wood ash contain potash and are good for the soil. Coal ash has no nutrients and is so fine that it spoils the soil's texture.

"My soil takes a long time to dry out in spring."

Keep off waterlogged soil or you will damage the structure. Water-retentive soil will require less watering in dry spells.

"What do the letters N-P-K on fertilizer packets stand for?"

These are the major nutrients – nitrogen (N), phosphates (P) and potash (potassium, K) – which all plants need in the correct quantities for healthy growth.

"I live in a dry part of the country. What can I do to cut down on watering in summer?"

Use mulches where you can, and add as much organic material to the soil as possible to make it more water retentive. Mulches should always be laid on damp soil.

"My soil flooded badly from a river last summer. What should I do with it now?"

Dig bare ground to break up any pans (hard areas). Most parts will need extra feeding because the fertilizers will have been leached (washed) out.

"There is a thin layer of soil over solid rock in my garden. How can I grow plants successfully?"

Plant in raised beds to give an adequate depth of soil.

"What are the white, thread-like things I sometimes find in the garden? They smell musty."

These are the mycelia of soil-borne fungi and often live within the roots of plants to the benefit of both.

"Is it necessary to test the soil regularly?"

It may be useful to test your soil occasionally for major elements, like nitrogen, and to check the pH, but if your plants are growing well, soil testing is generally not necessary.

"What does tilth mean?"

This is soil that has been broken down finely to make it suitable for seed-sowing.

"What is meant by planting 'in the green'?"

Some spring bulbs, such as snowdrops (*Galanthus*) and aconites, may take a while to start flowering again if you plant them as dry bulbs, so they should be dug up immediately after flowering but with the leaves still growing, then divided and replanted.

"What is naturalizing?"

This is when you plant bulbs permanently to come up in the same place, such as under trees or in lawns, every year.

"I have tried naturalizing tulip bulbs, but they never flower after the first spring. Why?"

Tulips need to be planted deep (aim for 15–20 cm (6–8 in) to the base) because the young buds may die in frozen ground.

"Is there anything I should bear in mind when naturalizing bulbs in grass?"

Make sure you plant to the correct depth for each bulb. Never cut off the foliage before it starts to turn yellow. Choose a slow-growing grass mix that will not choke the bulbs.

"What plants can I grow in my natural garden?"

Most daffodils, tulips, bluebells, crocuses, snowdrops (*Galanthus*) and fritillaries can be naturalized in grass.

"I forgot to plant any spring bulbs last autumn. What should I do?"

Most garden suppliers and many supermarkets sell them in pots in spring ready for planting. They are more expensive, but will give you an instant show.

"I've seen spring bulbs being sold off cheaply in midwinter. Are these a good buy?"

Yes. They will grow perfectly well, but the flowers may be a little later than normal.

"What happens if you forget to plant narcissi bulbs until the spring?"

They will still flower, but they'll produce fewer leaves. The following year, most bulbs will be blind (produce no flowers), but afterwards they should flower as normal.

"Can I plant hyacinths in the garden after they have been indoors?"

Yes, but the flowers will be smaller in subsequent years.

"My narcissi are dying and there are grubs in the bulbs."

These are the larvae of the narcissus bulb fly. Hoe around the plants as the foliage dies down to deter the adults from laying eggs.

"Why have none of my daffodil bulbs flowered this year?"

They may need feeding, or they could be short of water. Make sure you water them during a dry spring. Overcrowded bulbs will also stop flowering: dig them up and replant after flowering or in early autumn.

"I planted a lot of spring bulbs last autumn but none have come up. Why?"

Mice or squirrels might have eaten the bulbs. Try covering the soil surface with plastic netting, pegged down at the sides.

"Birds are pulling up my crocuses just as they come into flower. What can I do?"

This is difficult to prevent. Use small-mesh plastic netting on the surface as a barrier and move your bird table and feeders away from the area containing the bulbs.

"What do I do with the leaves of spring bulbs after flowering?"

Leave them for at least six weeks before cutting them off, as they feed the bulbs. Do not tie them up or loop them in elastic bands.

"Should I feed my daffodils?"

Yes. After flowering, give them one or two applications of a balanced fertilizer or fish, blood and bone before the leaves die down.

"The leaves of my daffodil bulbs are distorted and covered in yellow streaks. What is wrong?"

They have been infected with a virus that affects the bulbs' health. There is no cure, so dig them out and dispose of them, but not in the compost.

"Why do my crown imperials (*Fritillaria imperialis*) never flower after the first year?"

Sometimes rainwater enters the top of the bulb and rots the embryo flower bud. Plant the bulbs on their sides and make sure they are deep enough.

"Can you suggest bulbs to extend the season after the tulips have finished?"

Dutch iris, alliums, Persian buttercups (*Ranunculus asiaticus*) and spring-planted *Anemone coronaria* will all flower in early summer.

"Are there any spring bulbs that will flower in dry shade under trees?"

Grape hyacinths (*Muscari*) will flower almost anywhere. Naturalize them in big clumps for the best effect. Eventually, they will spread by seed as well as offsets.

"Tall spring bulbs look out of place in my tiny garden. What can I plant instead?"

Try snowdrops (*Galanthus nivalis*), chionodoxa, dwarf narcissi, such as 'Tête-à-tête', botanical tulips, such as 'Red Riding Hood', crocuses, scillas and *Allium moly*.

"I want to plant summer bedding plants in areas where there are bulbs. Can I dig up the bulbs?"

Yes. Dig them out carefully and plant them temporarily in a spare bit of ground. When the foliage has died down, the bulbs can be taken up, separated, cleaned and stored until autumn.

"My garden is full of aconites. How can I get rid of them?"

It would be a shame to do this because the foliage dies down early in summer so does not interfere with other plants. They also make good ground cover against weeds early in the season.

"Every year many of my bulbs rot soon after flowering."

Your soil may be too wet, or you may have planted rotten bulbs the previous autumn. Never plant soft bulbs, and, if your soil is soggy, try planting in ridges like professional growers do.

"Do naturalized daffodils seed themselves?"

If you do not deadhead them, they will self-seed. Thin, grassy leaves will appear around the parent plants in spring – these will eventually flower.

"How many years will spring-flowering bulbs in containers flower without repotting?"

It depends. In large pots and soil-based compost, they will flower for several years; crocuses, botanical tulips and narcissi should flower for at least three springs in multi-purpose compost.

"What are the white starry flowers that cover my border in mid-spring?"

These are the star of Bethlehem (*Ornithogalum umbellatum*), a close but hardy relation of the half-hardy chincherinchee (*O. thyrsoides*).

"I've planted Persian buttercups (*Ranunculus asiaticus*) in the flower border. Why do they never reappear after the first year?"

Plant in spring, not autumn, and lift the tubers after the leaves have died down.

"Many of my winter-flowering pansies are starting to look sick. Why is this?"

At this time of year, you will often find pansy aphid around the base of the plants. Spray well with an insecticide, or remove as many insects as you can by hand, and most will recover.

"How long do winter-flowering pansies last?"

If they are in a position out of direct hot sun, and if you deadhead them regularly and remember to feed and water them, they could last nearly all summer.

"Can I save spring bedding plants from year to year?"

Daisies (*Bellis*), pansies, sweet Williams (*Dianthus barbatus*), wallflowers (*Erysimum*) and some forget-me-nots (*Myosotis*) are short-lived perennials, and will last for a second year, but the flowering is always best the first year, so it is not worth it. Primulas and polyanthus can be divided to make more plants after flowering, and will last for many seasons.

"Is it worth saving the seedling forget-me-nots that come up in the garden?"

Generally, these will be a motley lot, but in an informal setting *Myosotis* seedlings can be allowed to grow up between other plants for an early flush.

"When should I sow wallflowers for next spring?"

Sow any time from late spring until early summer.

"I tried growing ornamental kale this winter, but they all ran to seed."

Kale will start to flower when the average winter temperature is above normal, which has happened in recent years. There is nothing you can do about it.

"I bought some Canterbury bells for my herbaceous border. Why did they die after flowering?"

This plant (*Campanula medium*) is a true biennial – that is, it grows one year and flowers, seeds and dies the next, so this was to be expected.

"What do I look for in a good bedding plant?"

Choose bushy specimens with healthy-looking leaves. Avoid those in full flower and with many roots coming out of the container base.

"When is the right time to plant out summer bedding plants?"

Wait until all risk of frost has passed, usually late spring. Retail outlets often sell bedding plants ready for planting out far too early.

"I have an unheated greenhouse but want to grow my own bedding plants. How can I get a head start with my planting?"

Buy plug plants from seed companies and young plant specialists. These are usually dispatched for growing on when no extra heat is needed in the greenhouse.

"I have no luck raising *Begonia semperflorens* from seed. Why do the plants fall over and die?"

You are sowing too thickly. Mix fine seed with silver sand to distribute it better. Over-high temperatures and poor light after germination can also be a cause, so do not sow too soon.

"What is a hardy annual?"

This is a plant that germinates, grows, flowers, seeds and dies in one season, and that can be sown outdoors in spring.

"What is the cheapest way of getting a quick showy display in summer?"

Use hardy annuals. Sown in mid-spring, they will be in flower by midsummer. Most will produce a second flush if deadheaded after flowering.

"Is there any way I can grow bedding plants without a greenhouse or cold frame?"

Some bedding plants, such as tobacco plants (*Nicotiana*) and French and African marigolds (*Tagetes*) can be sown outdoors in situ in early summer to give a good display in late summer and autumn.

"Is it worth buying end-of-season bedding plants?"

If they are stunted and root-bound, it is unlikely they will make good plants; otherwise, if you want to fill up some gaps, they are worth a try.

"Can I save my own seed from bedding plants?"

Many will not come true to their parents, but it is always interesting to see what comes up.

"What should I use to feed bedding plants?"

A weekly feed with a high-potash liquid fertilizer (such as tomato feed) may be necessary for containerized bedding plants; in good soil, they require little extra feeding.

"Can you suggest bedding plants to fill gaps in an herbaceous border?"

Choose plants that combine well with herbaceous perennials. *Antirrhinum*, ten-week stocks (*Matthiola incana*), tobacco plants (*Nicotiana*), cosmos and *Schizanthus* are among the best.

"What is a dot plant?"

This is a bedding plant that is taller than the rest and is often used to give accent and emphasis. Abutilons, cannas and *Ricinus* (castoral plant) are often used.

"Can you suggest some scented hardy annuals?"

Mignonettes (*Reseda*) have a lovely fragrance but are not much to look at, so surround them with more arresting plants. The flowers of night-scented stock (*Matthiola longipetala* subsp. *bicornis*) open at night and the scent is heavenly.

"What bedding plants will grow in light shade?"

Begonia semperflorens, busy lizzie (*Impatiens*), Mimulus, pansies, violas, lobelia and coleus (*Solenostemon*) will stand shade if it is not too heavy and dry.

"Which bedding plants will withstand hot, sunny conditions?"

Try *Mesembryanthemums*, pelargoniums, salvias, zinnias and *Portulaca*.

"Last year my hardy annuals grew tall and lush, but produced hardly any flowers. What did I do wrong?"

Either the soil was too rich or you did not thin out the seedlings enough. Thin out to the spacings specified on the seed packet and do not apply extra fertilizer.

"I want to grow some annuals on a trellis as a temporary screen. What would do best?"

Sweet peas (*Lathyrus odoratus*) are the obvious choice, but you could also try trailing nasturtiums, morning glory (*Ipomoea tricolor*) or black-eyed Susans (*Thunbergia*).

"Last year I saw some bedding plants that looked rather like conifers. What were these?"

These would be burning bush (*Bassia scoparia*). They turn bright red in autumn and make good temporary screens and dot plants.

"What is carpet bedding?"

This is a bedding scheme consisting of low-growing plants, often succulent or with interesting foliage, in a complex pattern. It can be used to good effect in front or small gardens, but watch out as they require a lot of attention.

"What is the best thing to do with bedding plants at the end of the season?"

Most are not worth saving and can be put on the compost heap, although one or two, like fuchsias, can be saved for another year.

"What is the difference between hybrid tea and floribunda roses?"

Floribunda roses have heads with many flowers; hybrid teas have a single flower or just a few blooms on each stem.

"What do the terms large-flowered and cluster-flowered mean?"

These are the new names for hybrid tea and floribunda roses.

"Is spring too late to prune?"

If the new shoots are long, it is better just to remove dead and dying wood and prune harder from late autumn to early spring next time.

"When should I feed my roses?"

Feed with a specific rose fertilizer in early spring, and again after the first flush of flowers in early to midsummer.

"Can you recommend some foolproof bush roses?"

The hybrid tea roses 'Silver Jubilee' (pink), and 'Royal William' (crimson), and the floribunda roses 'Sweet Dream' (apricot) and 'Iced Ginger' (ivory/copper pink) are healthy growers with excellent disease resistance.

"How long can I expect a rose bush to live?"

This depends on the cultivar and how well it is looked after, but a period in excess of 30 years is not unusual.

"Can you suggest a fragrant, strong-growing climbing rose with good disease resistance?"

You can do no better than the apricot-pink 'Compassion'.

"My climbing roses flower at the top of their trellises. How can I make them flower lower down?"

Untie them completely, pull down all the stems until they are as close to horizontal as possible, and then retie. Shorter, flowering shoots will soon form along the stems.

"What is the difference between a climbing rose and a rambler?"

A rambler usually flowers only once in a season, then produces the flowering wood for the following year. A climbing rose flowers on wood produced in the current season.

"How should I prune and train a rambling rose?"

Cut out all the shoots that have produced flowers immediately after flowering. Tie in new growths regularly as they appear.

"I have many established shrub roses that have become rather untidy. How can I tidy them up?"

Shrub roses can last for many years without regular pruning, but when they start to get out of hand, cut them back hard into the old wood between late autumn and early spring and feed well.

"Can you recommend a strongly fragrant rose for my partially sighted mother?"

A hybrid tea like 'Double Delight' (cream and red) or 'Deep Secret' (crimson) should fit the bill.

"How do you prune a miniature rose?"

Clip over the plant during its dormancy with a pair of sharp shears.

"The leaves of my roses have deep bites round the edges. What has caused this?"

This damage is caused by the leaf-cutter bee. It will not harm your plants, so there is no need to take action.

"Can I underplant my roses with ground-cover plants?"

It is possible, but these plants will compete with the roses for food and water, so you may find the roses do not thrive as well.

"Do ground-cover roses help with controlling weeds?"

Ground-cover roses are really only roses with a lax habit that makes them suitable for training over the soil. Weeds can still find their way through the shoots.

"I have no bare earth, but I want to grow a climbing rose. Can I plant one in a container?"

You need as big a container as you can manage and good-quality soil-based compost. Top-dress each spring and apply a good rose fertilizer; keep well watered, and give a weekly liquid feed every week during the growing season.

"My roses were covered with black spot last year. How can I prevent this happening again?"

Spray with a fungicide as soon as you see new growth in spring: with warmer winters, this may be as soon as mid-spring. Thereafter, spray monthly until the autumn.

"Can you suggest a shrub I can give as a gift for a golden wedding anniversary?"

Try *Choisya ternata* 'Sundance', the golden form of Mexican orange blossom. It is a good all-round shrub, as it is evergreen, and has fragrant flowers in summer. It can be trimmed to size in a small garden.

"I have a camellia in the conservatory that is 60 cm (24 in) high. When will it be safe to plant it out?"

Wait until there is no chance of frost. In future years it can be left outdoors, because camellias are reasonably hardy.

"Can you recommend some showy, flowering shrubs from spring to autumn?"

Forsythia 'Arnold Dwarf', *Deutzia* × *hybrida* 'Mont Rose', *Philadelphus* × *lemoinei* 'Lemoinei', *Buddleja* 'Lochinch', *Hibiscus syriacus* and *Ceanothus* × *delileanus* 'Gloire de Versailles' should see you through.

"Can you recommend a hedge that will attract wildlife?"

Pyracantha (firethorn) flowers attract bees and other beneficial insects in late spring and early summer, and birds will flock to the berries later in the year.

"Why does my *Pyracantha* never flower or have berries?"

You are pruning too hard and at the wrong time. Put away your secateurs and flowers will be produced from next year onwards. In future, remove new shoots in autumn, after the berries have set.

"Can you recommend an interesting screen for my back garden?"

Make a hedge by alternating forsythia and flowering currant (*Ribes sanguineum*), planting them 75 cm (30 in) apart. These will look beautiful when little else is happening in the garden, and the plants can be clipped regularly afterwards.

"Now that x Cupressocyparis leylandii (Leyland cypress) has become so unpopular, can you suggest another conifer that will make a good hedge?"

Western red cedar (*Thuja plicata*) is slower growing but will make a dense hedge reasonably quickly. The cultivar 'Zebrina' makes an attractive, variegated hedge. *Thuja* emits a lovely pine-like smell when brushed against or clipped.

"What can I use to replace an overgrown Leyland cypress hedge?"

Beech (*Fagus*), hornbeam (*Carpinus*), holly (*Ilex*), yew (*Taxus*) and laurel (*Prunus lusitanicus* or *P. laurocerasus*) will give you privacy, require less work and should not get out of control in the same way as Leyland cypress.

"I live near the sea and would like a flowering hedge. What will survive here?"

Escallonia is a good seaside hedge. It is evergreen and has flowers in white, pink or red, according to cultivar.

"Vandals regularly destroy the plants in my front garden. How can I beat them?"

Plant prickly shrubs. *Rosa rugosa*, *Berberis*, sea buckthorn (*Hippophae rhamnoides*) and ornamental blackberry (*Rubus*) will soon put a stop to this.

"How do I deal with a mossy lawn?"

A moss killer or lawn sand applied in spring will deal with the problem temporarily, but moss is often a sign that you are mowing too closely, so try raising the blades. If your lawn is in total shade, it is virtually impossible to rid of moss and algae, so if this is the problem, it might be better to replace the lawn with paving or suitable shrubs.

"What can I do? I need more parking space but don't want to lose my front lawn."

Honeycombed blocks can be laid like paving, then filled with fine soil and seeded. This makes a hard parking surface, which, once established, looks like a lawn and is maintained in the same way.

"I saw some flowering lawn seed the other day. Would this make a pretty feature?"

This is a mixture of lawn grasses and low-growing wildflowers that will still flower if mowed regularly. In a large, wild area, it is good for pathways through taller plants, but it can look untidy in a domestic garden.

"My lawn is more than half weeds. What should I do?"

Improving it may cost less and involve less hard work than killing the whole lawn off to get rid of the weeds, digging it thoroughly, then reseeding or returfing the area.

"Are grass paths practical?"

Only if they do not get much use. Alternatively, lay plastic reinforcing mesh before seeding or turfing.

"I have a lot of naturalized bulbs in my lawn. When can I start mowing?"

You will have to wait until six weeks after the flowers have died. The area will look rather pale and rough after the first cut, but will soon recover and blend in.

"Should I remove the grass clippings from my lawn?"

Left to rot, the clippings will kill some of the finer grasses and build up a waterproof thatch on the soil surface, so if you have a good lawn, it is best to collect them.

"What can I do about the dead patches that appearing on my lawn where my dog has urinated?"

This is actually a fertilizer overdose, and the patches should eventually grow over. Keep a full watering can handy to dilute the urine if you catch them at it, and train them to use a paved area instead.

"What is the difference between 'front-garden' and 'back-garden' lawn seed?"

Front-garden mixtures contain seeds of fine grasses that will withstand only moderate use. Back-garden lawn seed contains hard-wearing grasses capable of family use, but it looks less sumptuous.

"Is it possible to have a chamomile lawn?"

Lawn chamomile (*Chamaemelum nobile*) takes a long time to establish and must be weeded regularly until the plants knit together. The plants are comparatively expensive and not as hard-wearing as grass. Try a small area first to see how you get on.

"I spot-treated my weeds with glyphosate and the lawn is covered in dead patches. What can I do?"

Rake off the dead grass, fork over the patch to expose the soil, then reseed, using a lawn patch repair kit.

"When I make a new lawn, is seed or turf better?"

Seed is less expensive and easier to lay, but takes several months before it is able to cope with family use. Turfing is hard work and costs much more, but can be used in a matter of a few weeks.

"Do I have to use a fertilizer containing lawn weedkiller every year?"

No, once you have reduced the weeds to a manageable number, use a fertilizer-only product and spot-treat weeds as they appear.

"How do I deal with the cracks that appear in my lawn in dry weather?"

Water the lawn, then brush a mixture of horticultural sharp sand, fine soil and peat into the cracks. Large cracks may need to be over-sown with lawn seed.

"My lawn has been cut several times since winter. Why is it now pale and seems to have stopped growing?"

It needs feeding. Apply a slow-release lawn food, or, for a quick green-up, apply a liquid feed every three or four weeks.

"What do I do with suckers from next door's cherry tree that are appearing in my lawn?"

Cut the turf along the line of the suckers and peel back. Remove the roots bearing the suckers, replace the turf, then fill the cracks with sieved soil.

"My lawn is bumpy. I tried rolling it, but it made no difference. What can I do?"

Peel back the turf over the bumps, remove the excess soil, then replace and firm down all the turf. Fill cracks with finely sifted soil.

"How do I repair a broken lawn edge?"

Cut out the piece of turf with the broken edge and move it forward so the damaged part sticks out into the bed. Trim it level with the sound edge, then fill the hole in the lawn at the back of the turf with soil and reseed.

"How do I deal with hollows in my lawn?"

Top-dressing the area, using a little mixture at a time, will fill in most hollows after a while. Really deep ones should be filled with soil and reseeded.

"Why are there always lots of starlings feeding on my lawn?"

They are feeding on leatherjackets, the larvae of the daddy-long-legs. Working in sections, water the lawn at night and cover it with black polythene. Next morning, sweep up the leatherjackets that have come to the surface and give them to the birds.

"What can I do about the worm casts that are ruining my lawn?"

Wait till the lawn is dry and then scatter them thoroughly before mowing. Always pick up grass mowings and other organic rubbish.

"Yellow and dead patches have appeared where the postman takes a short cut across my front lawn. What can I do?"

This is fusarium (snow mould/mold), a fungal disease that is hard to remedy. Cut back on summer fertilizers and ask the postman to keep off the grass during frost. Leave reseeding dead patches until you are sure the disease has gone.

"Where do lawn weeds come from?"

Most are borne on the wind, some are spread by birds, human and animal feet, and some travel from weedy paths and borders.

"How can I get rid of a low, clover-like plant with yellow flowers that is growing in my lawn? It does not respond to weedkillers."

This is yellow suckling clover or lesser yellow trefoil (*Trifolium dubium*). It spreads by seeds and runners, so pull up or dig out as much as possible. Lower the mower blades to remove the flowerheads and repeat weedkiller treatments regularly.

"How short should I mow the lawn?"

A utility (hard-wearing) lawn should be cut to about 2.5 cm (1 in) in summer, and a luxury lawn to about 1 cm (½ in). Always leave the grass longer in spring and autumn and in times of drought.

"Why do my lawn grasses start to flower in late spring?"

This is because annual meadow grass has gained a hold. Feed regularly so the flower stalks can be mowed off; this will prevent this weed grass from spreading.

"Is a rotary mower better than a cylinder one?"

A cylinder mower will always give a finer finish, but modern rotary mowers with a grass box and roller can produce a finish that is nearly as good.

"My grass always seems to lie in the same direction, which spoils the appearance."

Mow the lawn in a different direction every time: up and down, side to side, and diagonally.

"Are home-grown vegetables really worth the bother?"

You do not really save money by growing your own vegetables, but you have the advantage of choosing varieties for taste and it is easy to grow them without pesticides.

"I always end up with a glut of most vegetables. How can I prevent this?"

Avoid the temptation of sowing the whole packet of seeds at once and sow successively, leaving a few weeks between sowings.

"How do I get rid of horsetail in my vegetable garden?"

Horsetail (*Equisetum*) has a deep root system and does not respond well to chemical weedkillers. However, it does not like being chopped off regularly, so if you hoe the tops off as soon as they grow above the ground, the plants will eventually get weaker.

"I am new to vegetable gardening. What crops should I start with?"

Start with easy crops: lettuce, spring onions (scallions), carrots, turnips, peas and broad (fava) beans, and take it from there.

"I grow my own carrots. Why do they always seem to have forked roots?"

Long-rooted carrots need deep soil without large lumps of fresh manure or compost in it. If you have added manure or compost recently, try growing a different crop first, like cabbage. If your soil is stony, you will need to work it thoroughly to remove as many stones as possible before you sow. Stump-rooted and round carrots are best to grow in shallow soils over rock or gravel.

"My garden is small and sheltered by a big tree next door. What vegetables should I try to grow?"

Most seed companies offer varieties of salad and root vegetables that can be grown in containers on the patio. Many suppliers offer tomato, courgette (zucchini), aubergine (eggplant) and pepper plants for sale in spring that can be grown outdoors, and these are as easy to grow on as summer bedding plants.

"My onions are always small and do not store well. What am I doing wrong?"

Feed with a high-nitrogen fertilizer until the bulbs have reached a reasonable size, then change to a high-potash feed.

"How do I know when it is warm enough to sow vegetables?"

Measure the soil temperature. Most vegetable seed will not germinate at a soil temperature less than 10°C (50°F).

"How can I encourage crop growths?"

Encourage the soil to warm up quickly by covering it with black polythene, and cover early crops with fleece or cloches.

"What is meant by chitting potatoes?"

This is when you expose seed potatoes to light in spring to encourage them to produce shoots. This gets the crop off to a good start.

"How deep should I sow vegetable seed?"

As a rule, smaller seeds, such as those for lettuce and carrots, should be sown at a depth of about 1 cm (½ in), whereas larger seeds need to be about 2.5 cm (1 in) deep. Always check the packet for guidance on both depth and spacing.

"What is a drill?"

This is the shallow channel into which seeds are sown.

"Last year, all my spinach ran to seed. How can I prevent it this time?"

Spinach needs to be grown quickly in moist soil and thinned out so the plants do not crowd each other. Keep the bed well watered and use the thinnings as salad leaves.

"Is there an easy alternative to spinach?"

Try leaf or spinach beet, which will survive less ideal growing conditions. When cooked like spinach, it tastes quite similar.

"When can I start cutting asparagus spears in my new bed?"

Never cut the first year after planting. In years two and three, you can cut spears sparingly; thereafter, cut as required.

"I still have a lot of leeks left over from the winter, but I want the ground for something else. What should I do?"

Dig them up carefully and heel them in (replant them temporarily) in a hole in a vacant part of the garden. They will last for several weeks like this, but will need more cleaning before cooking.

"When should I sow runner beans?"

If you sow them outdoors too soon, the seeds may rot or a late frost could kill the shoots. Either wait until late spring or sow the plants under glass in mid-spring and plant them out when they are about 8 cm (3 in) tall.

"I want to grow some asparagus. How do I start?"

Dig the bed thoroughly. Add plenty of well-rotted manure and a balanced fertilizer about a week before planting. Buy one-year-old crowns of an all-male hybrid, such as 'F1 Franklin'.

"Why have I never been able to grow good radishes?"

Radishes need to be sown little and often, grown quickly in good, moist soil, thinned out so the roots can develop, and picked when young and sweet.

"How deep should I plant asparagus crowns?"

Plant the crowns in trenches, 20 cm (8 in) deep and 30 cm (12 in) wide. To start with, cover the crowns with about 5 cm (2 in) of soil, and gradually fill in the trench as the shoots grow.

"I never have any luck with aubergines (eggplants). What am I doing wrong?"

Aubergines need a long growing season, so you should start the plants off early in a greenhouse or conservatory with some heat. If you're growing outdoors, give them a warm, sunny, sheltered spot.

"My rhubarb plants are spindly and produce few stalks. What can I do?"

Rhubarb needs an open position and deep, rich soil. Feed several times a season with a general fertilizer and keep well watered. Wait until the plants have recovered before pulling the stalks.

"What is crop rotation?"

This is a method of arranging the garden so that you never grow the same crop in the same ground in two successive seasons. The crop that follows should benefit from the previous one in some way. The purpose is to create healthy growing conditions and avoid a build-up of pests and diseases.

"What is green manure?"

This is a temporary crop of a quick-growing plant, such as rye grass or mustard, which is dug in while still green. It improves soil texture and adds some nutrients.

"Which vegetables could I grow in an ornamental border?"

Red-leaved lettuce, red cabbage, carrots, rhubarb chard, dwarf runner beans and cauliflowers with yellow or purple curds.

"I like to grow my own herbs but never have much success. Where am I going wrong?"

Most herbs like a free-draining soil and full sun. They also grow well in tubs and window boxes, so if your garden is shady or the soil is heavy, container cultivation may be the answer.

"As a child, I remember my uncle had a big parsley bed, but he never sowed new seed. My parsley only survives for one season. Why?"

Parsley is a biennial: that is, it produces flowers and seed in the second year and then dies. Your uncle's parsley bed probably kept going from year to year with self-sown seedlings.

"Why do I never have much luck growing mint?"

Mint likes cool, moist growing conditions. It is also best grown in pots because it can be invasive grown in open ground. Find a spot in half shade and never let the plants dry out.

"What compost or soil should I use to fill raised beds?"

Large raised beds can be filled with good topsoil with some well-rotted manure or garden compost added. Smaller beds are best filled with soil-based compost.

"Can I use multi-purpose compost in my raised beds?"

Yes, but you may have problems. It dries out quickly and generally needs replacing after every crop, so it is best to invest in soil-based compost from the start.

"There are worms in my raised beds. Will they do any harm?"

These may have come in on the roots of plants or in the soil if you have not used a sterilized compost, but they will do no harm at all.

"What should I use for the walls of my raised beds?"

Twin-wall, recycled PVCu boards are available online and from some seed companies. They take up little space, have good insulating properties and can be moved easily if necessary.

"A friend has used railway sleepers to make his raised beds. Is this wise?"

It is said that some of the wood preservative will leach out of the timber and damage the crops, but usually the sleepers have weathered sufficiently for this not to be a problem.

"What are the optimum dimensions for a raised bed?"

Your beds should be no more than 1.2 m (4 ft) wide and 3 m (about 10 ft) long, so that you have good access to the soil from different sides.

"What is the best depth for a raised bed?"

Salad leaves and spinach can be grown successfully in a soil depth of no more than about 15 cm (6 in), but most vegetables will do better in beds that are about 30 cm (12 in) deep.

"I would like to put some raised beds on my patio. Can I put them directly on the slabs?"

You may find that if they are there for more than one season they will stain the slabs, so cover the slabs with polythene first. You also need to check that the water can drain away properly.

"Last year I grew broad beans in raised beds, but they all flopped over. How can this be avoided?"

They will need to be supported. Make a framework of thin canes, woven together so they stay upright, or use brushwood, pushed into the compost.

"How far apart should broad beans be spaced in raised beds?"

You should find that they will grow well if the plants are about 15 cm (6 in) apart.

"Can I grow tomatoes in a raised bed?"

They are difficult to support, but if you plant a hanging basket variety, such as 'Tumbler', around the edge of the bed, you should get a good crop. The middle area can be used for another vegetable.

"What varieties should I look for when I'm sowing vegetables in raised beds?"

Look for varieties described as 'patio' or 'baby' vegetables. These mature quickly and produce good-looking vegetables at closer spacings.

"Will (bell) peppers grow in raised beds?"

They should grow well if you choose types that are specifically named as 'patio' varieties. Space them about 30 cm (12 in) apart.

"Can I grow long-rooted beetroot in a raised bed? I find these are better for slicing."

When they are grown in a shallow raised bed, long-rooted beetroot will push themselves out of the compost. However, this does not seem to affect the crop, and you can cook and slice them as normal.

"Will kale grow in a raised bed?"

You will need a fairly large raised bed, but the variety 'Pentland Brigg' is worth trying. Sow the seed in late spring and plant them out in another bed in the second half of summer. Space them 38 cm (15 in) apart.

"Will garlic grow in a raised bed?"

Garlic does grow well in raised beds, but it needs a long growing season. Plant in early spring, or wait and plant in autumn.

"Could I grow courgettes in a raised bed?"

Raised beds are ideal for courgettes and marrows, because the plants can be allowed to hang over the sides. Space them at least 45 cm (18 in) apart for best results.

"Can I grow runner beans in raised beds?"

Climbing beans are tricky to grow in smaller raised beds because they are difficult to support adequately. Dwarf runner beans and French beans are excellent for this type of cultivation, though.

"I find the temptation is to grow vegetables in raised beds far too close together. How can I avoid this?"

Try growing a pinch of each type in a small pot or plastic cell. When these are large enough, they can be planted out at the correct spacings, and any thinning used as baby vegetables or in salads.

"Which is the best variety of lettuce to grow in a raised bed?"

'Little Gem' is possibly still the best, or you could grow a cut-and-come-again type like 'Salad Bowl'.

"Is it sensible to try growing potatoes in a raised bed?"

You will need a fairly large, deep bed because the tubers should be planted about 30 cm (12 in) apart and 12 cm (5 in) deep. Remember, too, that you will have to earth up the potatoes as they grow. Grow an early variety, like 'Swift', so you can use the bed for something else later in the season.

"What is the best material for the paths between raised beds?"

If you think you will keep the beds for a long time, gravel or paving slabs are the most practical. Otherwise, used coarse chipped bark.

"I am away from home a lot. How can I keep my raised beds watered adequately?"

A soaker hose and an electronic timer connected to an outside tap will take all the hard work out of keeping the beds damp.

"What should I use to feed my raised beds?"

Top-dress with a general fertilizer two weeks before sowing or planting in spring, then water with a soluble feed once or twice a week throughout the growing season (in addition to regular watering).

"How often should I change the compost in a raised bed?"

If you are using soil-based compost, you will not need to change it for several seasons, but it will require regular topping up, as you remove some soil every time you remove a plant.

"Can I grow fruit in raised beds?"

Most fruit bushes have too large a root system to grow well in raised beds, but strawberries and lowbush blueberries (bilberries) can be grown successfully.

"How can I stop birds digging in my raised beds and throwing the compost everywhere?"

Birds often dig in raised beds looking for insects and similar food. If it becomes a problem, you may need to make a frame and cover the beds with fleece or netting.

"Last year I grew parsnips, but the roots did not form. What am I doing wrong?"

It is essential that you thin parsnip seedlings in stages as soon as they are large enough to handle. If you do not do this, the roots will not have enough space to swell.

"What is a rootstock?"

All fruit trees, like apples, pears and plums, are grafted on to a different root, which modifies the growth. For instance, M27 is a dwarfing apple rootstock; 'Stella' is a semi-dwarfing cherry stock.

"I love apples and pears, but have a small plot. Can I have a fruit tree?"

'Family' fruit trees have two or more varieties on one tree, so will give you a mini-orchard. Otherwise, consider cordons, fan-trained and espalier trees.

"I bought an apple tree supposed to be suitable for a small garden, but it has grown enormous. Who is to blame?"

You may have planted it too deep. The kink in the stem near the base, which is where it was grafted, should be well clear of the soil.

"Can you recommend a really nice pear?"

Try 'Concorde'. It is self-fertile and produces huge crops early in its life.

"I have seen fruit trees for sale with the roots wrapped in elastic mesh. How do I plant these?"

The supplier will probably tell you to leave the net on, but if the tree is dormant it will establish better if you cut it off and spread out the roots.

"What is meant by self-fertile?"

Many fruit trees need the pollen from another variety to set a crop. A self-fertile tree will produce a crop without the need for cross-pollination.

"Will an apple tree pollinate a pear?"

No. Apples will only pollinate apples, pears only pears, and so on.

"Is there a fruit tree or bush that will grow against a shady wall?"

The cooking cherry 'Morello' can be grown in shade. Blackberries and blackberry hybrids (such as the loganberry or tayberry) will also tolerate all but the densest shade.

"Birds always eat all my cherries. How can I stop them?"

It is impossible to prevent birds from getting the lion's share of fruit on a tall cherry tree. Try netting some lower branches before the fruit starts to ripen.

"When should I prune raspberries?"

Prune summer-fruiting raspberries after fruiting, and autumn-fruiting varieties in spring.

"Which raspberry canes should I prune?"

Remove those that have borne fruit completely, and also any short, weak canes. Tall canes can be cut back slightly in spring for easier access to the crop.

"All my old gooseberry bushes had mildew every year, so I have taken them out. Are any varieties more mildew resistant?"

The varieties 'Invicta' and 'Jubilee' are both resistant and produce huge crops.

"Is it necessary to cover strawberry beds with straw? I cannot get hold of any."

Straw is traditionally used to keep the fruit off the ground and protect it from dirt. You can plant through black polythene or landscape fabric just as effectively.

"I've let my blackcurrant bushes get out of hand. What should I do?"

Sacrifice the fruit on one or two bushes for a season by cutting them back to ground level. New shoots will appear over summer, and these will fruit next season.

"How do I prune a blackcurrant bush to keep it healthy?"

Remove about one-third of the old shoots every year. This will encourage new growth from low down and keep the bush young.

"I have grown some melon plants in the greenhouse. Can I plant them outdoors in summer?"

Most melon varieties will fruit satisfactorily only at higher temperatures, approximately 21°C (70°F), and need greenhouse cultivation. If you have a warm, sheltered spot in full sun, you could try one or two plants.

"I've grown a peach from a stone. Will I ever get any fruit?"

Productive peach trees grow well from stones. Train it on a sunny wall as a fan and it will, eventually, bear fruit.

"Can I grow an apple from a pip?"

Yes, but it will not come true to the apple variety the pip came from – and unless you bud it on to a dwarfing rootstock, it will make a big tree and take many years to fruit.

"Are 'patio' fruit trees worth growing?"

These are generally grafted on to dwarfing rootstocks and can be quite weak. You will get better results if you plant a tree on a normal rootstock in a large half-barrel filled with soil-based compost.

"How can I grow a quince?"

The quince makes an attractive, ornamental tree. Choose the variety 'Vranja' for a good crop of delicious fruit.

"My fig tree produces a lot of small fruits in late summer. Why do they never come to anything the following spring?"

Remove all embryo fruit after harvesting the figs, but retain any that are about the size of a pea, because these will start to develop the following spring.

"Why do I have to prune?"

It depends on the type of plant. Roses are pruned to encourage flower production, shrubs to keep them young and healthy, trees for shape, and so on.

"Do I have to prune shrubs every year?"

No. Young specimens may need pruning annually at first to encourage strong growth; after that, pruning is usually done when the branches get overcrowded or no new wood is produced.

"When do I prune a hydrangea?"

Never, if you can help it. To keep it a reasonable size, remove about one-third of the old shoots at ground level in spring every year to produce new shoots from the base. In this way, you will always have flowers, plus young wood that will flower the following year, and the whole bush will be rejuvenated in four years. If you cut back all the shoots, you will remove the flowering wood as well. However, if the plant is overgrown, you may find it better to cut it back to ground level, sacrifice the flowers for one season and then treat it annually, as described above.

"I have inherited some apple trees that have been hard pruned every year and do not produce much fruit. Can I rejuvenate them?"

Mature fruit trees need little pruning, other than the removal of weak, dead and crossing branches. Follow this rule and do not prune the tips. The trees will start to fruit again.

"My flowering cherry has grown over a flower border. When should I cut it back?"

Members of the cherry family, such as ornamental and fruiting cherries, plums, peaches, nectarines and the like, should be pruned during warmer times of year (late spring to early autumn) to avoid infection with fungal diseases.

"I have willows (*Salix*) and dogwoods (*Cornus*). When should I prune them?"

Cut them back hard every two or three years for the best results.

"When is the best time to prune a magnolia?"

Never! If you need to prune a magnolia you have chosen the wrong type. Cutting off shoots will make matters worse and affect flowering. Instead, remove the lower branches completely so you open up the space beneath.

"Some years ago, I planted a blue cedar. It is getting big. How and when can I prune it?"

It is difficult to prune some conifers, such as *Cedrus atlantica f. glauca*, without spoiling the shape. You might have to think about removing it and replacing it with something more suitable.

"How far back can I cut a eucalyptus?"

Eucalyptus benefits from hard pruning in spring and will shoot up again, even if it's cut to the ground.

"I have a *Cordyline australis* with a 6 m (20 ft) bare trunk. It looks ridiculous. Can I do anything with it?"

In warmer areas, and with a mature specimen like this, it may be possible to shorten the trunk. Rosettes of leaves will then form lower down.

"What happens if you cut an old rose bush back really hard?"

If it is healthy, it might produce new shoots. Otherwise, it probably won't recover. If you are fond of it, treat it gently.

"My English lavender bush has got leggy. How can I improve it?"

You must remember to trim lavender regularly. It should be pruned every year after flowering by removing most of the current season's soft shoots, but it will not regrow from woody branches.

"My laurel hedge has grown too large. Can I save it?"

Laurel (*Prunus lusitanicus* and *P. laurocerasus*) recovers well from hard pruning in spring, but it may look a bit bare the first year.

"The man who helps me in my garden has pruned my yew trees back really hard. Will they recover?"

Yew (*Taxus*) is one of the few conifers that can be cut back into really old wood. They will sprout again and be better for it.

"When should I trim winter-flowering heathers?"

Clip after flowering every spring, but do not cut back too hard or they will not recover.

"I have a hedge of laurustinus (*Viburnum tinus*), which I trim in autumn every year. Why does it never produce flowers?"

You are trimming off all the flowering shoots. In future, trim in spring.

"When should I prune a rosemary bush?"

The best time is in late spring or early summer, so it has a chance to recover before the frosts.

"I have a bougainvillea in my conservatory that is getting out-of-hand. Can I prune it?"

Shorten all the shoots back to the supports. It will start producing bracts on new wood later in the season.

"My forsythia is beautiful every year. I never prune it. Should I?"

It's not essential, but if you remove a few old branches after flowering every year, the bush will live longer.

"Should I prune my hardy fuchsias back in spring?"

Pruning will keep the bushes young and make them a more manageable size, but it is not vital.

"What are the red spots that are spoiling my spring daisies (*Bellis perennis*)?"

This is a fungal disease called rust. Apply a fungicidal spray as soon as you notice the problem, and the new leaves and flowers will be healthy.

"Sparrows are tearing my yellow polyanthus to bits. How can I stop this?"

Push twiggy sticks in between the plants to protect them.

"Last year my forget-me-nots (*Myosotis*) were ruined by a white, powdery coating. How can I stop this happening again?"

This was powdery mildew, a fungal disease, which can be contained but not cured by applying fungicide. Spray before the disease starts to appear.

"Why do my pulmonarias get covered with a white mildew and then turn brown and look unsightly?"

Pulmonarias are susceptible to powdery mildew. Spray them with a fungicide soon after the new leaves appear in spring. Do not wait for the disease to appear.

"What's wrong with my hebes? They look sick and some have died."

They have probably been affected by the fungal disease mildew. Trim the live shoots back lightly and spray immediately with a fungicide. You may need to repeat this at two- to three-week intervals throughout the summer.

"Some of the branches of my spotted laurel (*Aucuba japonica*) have wilted and turned black. Shall I dig it up?"

This die-back is a problem with *Aucuba*, but if you remove the affected branches completely, the rest of the shrub will usually recover.

"My Pyracantha (firethorn) is covered in white fluff. What is it?"

This is woolly aphid, which is difficult to deal with, as it is waterproof and cannot be contained with contact insecticides. A systemic insecticide will kill woolly aphid temporarily, but washing it off with a strong jet of water is just as effective and lasts just as long.

"My pyracantha is covered in black spots and the buds have shrivelled. What is the matter with it?"

This sounds like a bad infection of scab. As it has got to this stage, it is best to remove the shrub completely, wait a year or so, and replace it with a scab-resistant cultivar such as 'Saphyr Rouge'.

"The needles on my yew (Taxus) look yellow and sick. What is wrong?"

This is caused by conifer spinning mites, which suck the sap out of the leaves. There are no good controls, but spraying with a generic insecticide every three weeks from mid-spring may help.

"Every year, the leaves of my flowering almond are blistered and distorted. What causes this?"

This is a fungal disease called peach leaf curl. Spray with Dithane 945 in late winter, then again three weeks later and before leaf fall in autumn. Remember to clear away all fallen leaves as soon as you see them, and feed and mulch around the tree.

"Some nasty white grubs have eaten the roots of my overwintering fuchsias in the greenhouse. What are these and how can I stop the same thing happening again?"

These are vine weevil grubs, now a huge problem in gardens. There is one chemical that can be watered on the compost, which will control these pests for up to six months. If you prefer to grow organically, biological controls are available online. Using soil-based compost, not overwatering and putting a mulch of grit or pea shingle on the pots will deter the adult weevils from laying eggs.

"My Judas tree is dying back. The dead wood has orange spots on it."

This is coral spot, a fungal disease. It mainly affects dead branches, but in some shrubs, like *Cercis siliquastrum*, it can spread into live wood and kill it. Remove all affected parts completely as soon as you see them.

"Where can I obtain ladybird and lacewing chambers?"

Some larger garden suppliers stock these. Organic gardening suppliers usually have them in their brochures, or you can find suppliers on the internet.

"A visiting cat uses my seedbeds as a lavatory. How can I scare him away?"

A number of homemade and proprietary cat repellents are available, including water sprays, but most of them are not terribly effective. The only certain way is to cover your seedbeds with fleece until the plants can look after themselves.

"What can I do about those horrid grubs that infest my apples every year? I do not want to use chemicals."

Use a codling moth trap. This contains a pheromone that is attractive to the male moths, which then get stuck on a sticky card inside the trap. Hang the trap in the branches in late spring and renew the pheromone after about six weeks. They are available online and from larger garden suppliers.

"Last year I had a wasp nest in the garden. What do I do if it happens again?"

Your local authority is the best organization to contact. They will either send a pest control officer to deal with it or recommend someone who can help. However, wasps are beneficial insects, as the larvae eat aphids, so unless someone in your household is allergic to wasp stings, it may be better to leave it alone.

"I have had greenfly on many of my plants since winter. I thought they died at the end of summer. Where do they come from?"

This has become more common as winters are getting warmer. Aphids are easily curbed by a number of organic or inorganic pesticides, but attracting beneficial insects into the garden by providing shelter for ladybirds, lacewings and the like may make pesticides unnecessary.

"Last year I had mildew on many of my plants. Is one plant infecting the others?"

No. These mildews are plant specific, so will not spread from one genus to another. Start spraying with a fungicide early in the season, before it appears, to prevent infection.

"Bees seem to be drilling holes in my lawn. Will they damage it?"

These are the nests of solitary bees. A healthy lawn should not be damaged, and the bees are unlikely to sting in normal circumstances.

"Ants are eating the roots of my alpines and killing them!"

Garden ants do not eat roots or other plant material, but they will disturb the soil so some plants may die of drought. Control aphids, because ants 'farm' them for the sweet honeydew they secrete.

"What is nibbling the edges of the leaves of my peas and broad (fava) beans?"

This damage is caused by the pea and bean weevil. Usually the problem is not serious enough to make spraying with an insecticide necessary.

"The tops of my broad beans get full of blackfly every year."

Remove the shoots once the first beans start to set. Broad bean tops can be cooked as a delicious green vegetable.

"Last year all my onions had white rot. Is there anything I can treat the soil with before planting my sets this spring?"

Unfortunately not. You will need to grow your onions (and any ornamental alliums) in a completely different part of the garden for at least eight years.

"All my wallflowers have died off. When I pulled the plants up, the roots looked like dahlia tubers. What caused this?"

This is clubroot disease, which affects all members of the cabbage (*Brassicaceae*) family, including ornamental ones like wallflowers and stocks. Do not grow similar plants in the same place for several years, and do not compost affected material.

"What has caused the little, tortoise-like swellings all over the branches of my oleander?"

Scale insects affect many outdoor and conservatory plants. Where the affected plant is growing in a pot, watering with a vine weevil killer will control it for three months or more. The remains can be removed with an old toothbrush.

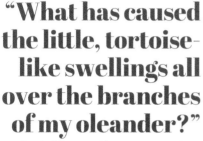

"My friend grows organically and recommends the liquid from boiled rhubarb leaves as an insecticide. Is this safe?"

The liquid is highly toxic, killing both beneficial and non-beneficial insects alike. And if it's sprayed on food crops, it could make you ill.

"I do not like using chemicals, but my vegetable garden has just about every pest and disease there is every year. What can I do?"

Keep all your crops covered with fleece or other plant protection sheeting from sowing to harvesting as a barrier to all manner of nasties.

"Every year my carrots are rendered inedible by little white maggots in the roots. What should I do?"

Choose a carrot-fly resistant variety, like F1 'Flyaway'. Carrots in raised beds are less likely to be affected because the adult flies home in at ground level.

"Garden chemicals are so expensive. My brother is a farmer – can I use his chemicals in the garden?"

In some countries it is illegal for an amateur gardener to use agricultural chemicals, and also illegal for a professional grower to supply them or decant them into any container not intended for the relevant product. It is best to check with your local authority.

"I sowed a row of early peas a few weeks ago, but nothing has appeared. Are they dead?"

Mice and birds will eat the seed almost as soon as you sow it. Cover immediately with fleece or another transparent crop-protection material.

"How can I keep rabbits out of my vegetable patch?"

Surround your patch with wire netting, buried to a depth of at least 30 cm (12 in). To cut down on cost, divide the area into smaller beds and protect those containing crops that rabbits like most, such as lettuce, carrots, peas and spinach.

"I live in the country and deer are ruining my garden. How can I keep them out?"

The only effective way to control deer is by erecting high fences or netting. Some people recommend hanging unwashed human hair around the garden as a deterrent.

"How can I start making my own compost?"

Contact your local authority because many have schemes that offer compost bins at discounted prices. Start with two: one to fill and one in which to finish rotting.

"Why is my compost always a slimy, smelly mess?"

You are adding too much green, wet material. Alternate grass cuttings and kitchen waste with shredded paper and chopped-up prunings, and the situation will soon improve.

"My compost heap doesn't seem to be breaking down. Can you recommend a good, cheap composting agent?"

Accelerators should not be necessary if you add the right ingredients in the right proportions, so layer coarser materials with green waste, such as grass mowings. Urine is an excellent accelerator if you are not squeamish.

"Can I put poisonous plants, like rhubarb leaves and laburnum, in my compost bin?"

Yes. Poisonous and non-poisonous plants break down during composting in exactly the same way to give a material that is safe for both other plants and feeding vegetables.

"Can I put paper in the compost?"

Yes, if it is finely shredded and mixed with wetter material, such as comfrey leaves or grass clippings.

"Will dog hair and the contents of the vacuum cleaner break down in the compost heap?"

Yes, but generally more slowly than plant remains. Add hair in small quantities. Carpet fluff must be from rugs and carpets of organic origin, such as cotton and wool.

"How do I keep rats and mice out of my compost bin?"

Stand the bin on concrete to prevent them digging through from the soil. However, they may chew through plastic, so block up any holes with wood as soon as you can.

"Should a compost bin be green or black?"

Dark shades absorb the sun's heat better than light ones, so the compost will heat up and break down more quickly.

"Is a square bin better than a round one?"

If you have more than one bin, it is better to use square ones, which can stand against each other to conserve space and keep each other warm.

"Does a compost bin need air holes?"

As compost is produced mainly by bacteria that work without the presence of air, ventilation is unnecessary and will slow down the rotting process.

"What is meant by 'turning the compost'?"

If you make compost in a heap rather than a bin, you need to turn the material periodically so the unrotted outer layer is turned to the inside to break down. This is unnecessary if you are making compost in a bin.

"Is it true that you can make compost in a polythene bag?"

It is quite possible to make good compost by filling a heavy-duty, black polythene bin bag with soft, well-shredded green waste. Tie the top tightly and leave it in full sun until the compost is formed.

"My compost is rather coarse. Can I still use it?"

It sounds as though it needs more time to rot. If you want to use it immediately, put it at the bottom of planting holes or runner bean trenches to finish off.

"I've seen slugs and snails in my compost bin. What should I do?"

If they cannot escape, they will help to break down the compost, but if you intend to empty the bin, just add a few slug pellets to the surface of the material a week or so beforehand.

"Should I add water to my compost bin?"

If you have enough green material, water should not be needed, but if the compost looks dry and has stopped rotting, a little warm water may get it going again.

"Can I put cooked kitchen waste in the compost bin?"

No, it will attract vermin and smell terrible as it rots. Special kitchen-waste rotters, which process cooked meat, vegetable waste and dairy products, are available from organic appliance suppliers.

"Do I have to add worms to my compost bin?"

Worms will usually arrive on their own in the soil around plant roots.

"There are no worms in my compost bin. Does this matter?"

Worms help to break down plant material into good compost, but if you add a good mixture of different waste, you will usually get a high-quality result anyway, although it might take a little longer.

"Do compost tumblers really work?"

It depends what you put in them. They will break down finely chopped vegetable and fruit parings reasonably well, but they need to be turned at least daily, and this can be hard work as the container fills up.

"I have thought of buying a wormery. Will the results justify the cost?"

If a wormery is used strictly according to instructions, it makes good dry material as well as an effective liquid feed. However, if anything goes wrong (if you add the wrong materials, or if the bin gets too hot, too cold, too dry or too wet) the worms die and the wormery stops working.

"Someone told me that I should add soil to my compost bin. Is this true?"

In a traditional compost heap, soil is sometimes layered with plant rubbish to add micro-organisms and retain heat in the rotting material. This is not necessary with a modern bin.

summer

"What is meant by 'half-hardy' on the labels of some plants?"

This is the term used to describe plants that will grow happily outdoors during the summer, but cannot tolerate frost and will need the protection of a slightly heated greenhouse in winter. Many patio plants, like *Brugmansia*, lantana and citrus, fall into this category.

"I cannot keep my perennial wallflowers alive for more than a year or two. Where am I going wrong?"

They are naturally short-lived plants but are easily propagated from cuttings in summer.

"Can you suggest some late-flowering perennials to brighten my border?"

Try helianthemums, Michaelmas daisies, *Schizostylis*, *Sedum spectabile* and *Anemone* × *hybrida*.

"What should I do with lupins and delphiniums after they have finished flowering?"

Cut them down almost to soil level. They may produce a second flower flush later in the season.

"My flower border is dry. Are there any perennials that will do well here?"

First, dig in plenty of organic material to help hold water. *Achillea*, *Echinops* (globe thistle), *Kniphofia* (red hot poker), nerines and *Stachys* are just a few perennials that should thrive in these conditions.

"My garden is in permanent shade from the house next door. Are perennials out of the question?"

No. Try *Alchemilla mollis*, *Darmera peltata*, *Caltha*, *Trollius* and hostas for an attractive effect.

"Some nurseries recommend hostas for dry situations, some for moist. Which is correct?"

Hostas are, in fact, pretty tough and will withstand a wide range of conditions.

"How can I stop my flower border always looking a mess because the plants flop over?"

Provide supports before the plants have grown too tall. They will never look right if they are tied up after they have flopped.

"Herbaceous plant supports can be horrendously expensive. What is the alternative?"

Either use brushwood, pushed into the ground at the start of summer for the stems to grow through, or, if not available, ring the plants with canes and run soft string around them as the plants get taller.

"Last autumn, I bought some hardy perennial plug plants, which I potted on into 15 cm (6 in) pots. I overwintered them in an unheated greenhouse. When can I plant them out?"

Early summer is the best time, so they can get established before winter.

"My bearded irises have stopped flowering. Should I discard them?"

Take them up, discard the old, woody rhizomes in the middle of the clump, and replant the newest parts. Cut back any long leaves after replanting.

"I planted some bearded irises in my bog garden, but they rotted. I thought irises liked moist conditions?"

Not all do. Bearded irises like dry soil and a good baking in full sun. Plant *Iris siberica* in your bog garden.

"I have a sunny, moist patch at the bottom of my garden. What hardy perennials should I plant there?"

There are plenty that like these conditions. Try *Ligularia*, *Caltha*, *Filipendula*, *Euphorbia palustris*, *Omphalodes*, *Lythrum*, *Rodgersia* and *Lysimachia*.

"I live near the sea and think my soil contains some salt. What perennials will tolerate this?"

Choose those that have been bred from plants that naturally live near the coast, such as sea lavender (*Limonium*), perennial wallflower (*Erysimum*) and sea holly (*Eryngium*). Cupid's dart (*Catanache*) and globe thistle (*Echinops*) should also do well.

"I bought a pampas grass (*Cortaderia*) five years ago and it has never flowered. I have been told this is because I do not set fire to it every spring. Is this true?"

This is often recommended as a way of both tidying it up and giving it a dose of potash to help it flower at the same time, but it can also set fire to the roots, so cutting back is better.

It sounds as if yours is a poor seedling plant. Dig it up and replace it in late summer with one already showing plumes.

"I had some beautiful Himalayan poppies (*Meconopsis*), but this year they have disappeared."

Many of them die after flowering. Do not deadhead and you are likely to get seedlings the following year.

"Can you suggest a perennial to edge a rose border?"

Catmint (*Nepeta*) is often used for this. Cut it back regularly in summer to keep the plants neat and to encourage flowering.

"I had a lovely collection of heucheras. What could have eaten all the roots?"

This is possibly damage caused by vine weevil grubs. Dig the soil over thoroughly and pick out (and kill) any small, white grubs before replanting, then mulch around the plants with chippings or pea gravel.

"Is it a good idea to plant spring bulbs between perennials?"

It will extend the season for display, but you must be careful that the bulb leaves do not smother emerging plant shoots. They may also make the border look untidy after flowering until they die off.

"I'm gardening on a budget and can't afford to pay the prices most garden stores charge for hardy perennials. Is there an alternative?"

If you don't mind waiting, most of the common kinds are easily raised from seed, and some will even flower in their first season. Check the seed brochures for what is available.

"Could I make a perennial bed in the middle of my lawn?"

Yes, but remember that the bed will be seen from all sides, so put the tall perennials in the middle, not at the back. Choose sturdy, shorter-growing varieties.

"Can hardy perennials be mixed with other kinds of plants?"

Definitely. They can really brighten up a shrubbery or add interest to a rose bed.

"When should I split the perennials in my herbaceous border?"

The recommendation is every three years, but if they are doing well, there is no need to do this job until flowering and growth are affected. Autumn or early spring are the best time.

"How can I get rid of the bindweed growing through my herbaceous border?"

Wait until the bindweed (*Convolvulus*) has grown to about 30 cm (12 in) high, then paint the leaves with glyphosate. Several applications may be necessary.

"My herbaceous border is completely overrun with couch grass. Is there a chemical I can treat it with?"

Any chemical that will kill couch grass will also kill the plants in the border. The best way to get rid of this weed is to lift the plants in autumn, split them and pick out every bit of couch grass root. Dig over the border and remove all the roots from the border soil before you replant your perennials.

"I have a new herbaceous border and it looks sparse. Have I spaced the plants out too much?"

No, they need room to establish. Fill in with annuals and summer bulbs for temporary interest.

"Should I mulch an herbaceous border?"

Mulching in late spring or early summer will conserve moisture and deter weeds.

"Should I cut my perennials back at the end of summer?"

Most gardeners like to tidy the borders at the end of the season, but slightly tender perennials, like *Kniphofia* and penstemons, are best left till spring.

"What is the purpose of deadheading?"

It is done to encourage the rose bush to put all its energy into producing new flowering shoots for a second flush.

"What is the correct technique when deadheading?"

Strictly speaking, you should remove both the spent flowerhead and the stem below to the first leaf with five leaflets. In practice, the results are just as good if you just pinch out the dead flower.

"I've got sandy soil in my garden. Can I grow roses?"

Keep them well watered and mulched, and feed every month from late spring to early autumn with half the recommended application of a rose fertilizer.

"My old roses produce a lot of suckers. What should I do?"

Trace them back to the roots and pull them off. Do not cut them. Do not damage roots by deep digging.

"What are the best roses for pot pourri?"

Heavily scented, old-fashioned shrub roses and red, strongly scented hybrid teas, such as 'Josephine Bruce', 'Red Devil', 'Fragrant Cloud' and 'Deep Secret', will give good results.

"Can you suggest some climbing roses to grow up obelisks?"

Choose moderate-growing varieties with an upright habit, such as 'Aloha', 'Golden Showers', 'Joseph's Coat' and 'Handel'.

"Will a climbing rose grow on a north-facing fence?"

In this position, try 'Danse du Feu', 'Mme. Alfred Carrière' and 'Climbing Etoile de Hollande'.

"Is there a chemical-free way of preventing disease in roses?"

Choose new cultivars that have been bred for disease resistance. Space them well apart, keep the soil moist and feed them at least twice a season.

"What is meant by 'English' roses?"

This group has flowers that resemble the old shrub roses, but they are repeat flowering and can be formally pruned as floribundas or lightly tidied like old-fashioned shrub roses.

"Can I use grass clippings as a mulch on a rose bed?"

Yes. The mulch will keep the roots cool and moist, but if you have annual meadow grass in your lawn, you may get problems with seedlings from this practice.

"What kind of rose can be grown into an old apple tree?"

The strong-growing ramblers 'Wedding Day', *Rosa filipes* 'Kiftsgate' and 'Bobby James' are suitable, but they must be pruned every year after flowering or they will get out of hand.

"I planted some ground-cover roses in a narrow bed between my house and the drive, and they are spreading too far. How can I control them?"

Treat them as climbers. Put training wires or trellis on the wall and train them up this. They can be pruned as ramblers, or just clipped over lightly after flowering and again in spring.

"Is there an early-flowering rose that I can grow in my shrub border?"

One of the earliest roses to flower is *Rosa xanthina* 'Canary Bird', whose yellow flowers appear in late spring.

"I would like to grow roses with hips that I can make into wine."

The best for this is any cultivar of *Rosa rugosa*. Even if you are not a wine-maker, the hips will extend the season of interest.

"When and how should I prune a rose hedge?"

Prune after the first flush of flowers and again in autumn – unless the roses bear ornamental hips, in which case you should prune in winter or early spring.

"The head of my standard rose has died. Can I cut the stem back and start again?"

No. The stem is only a rootstock, and you will get wild-rose-type flowers from it.

"I have a 'Dorothy Perkins' rambler trained as a weeping standard. How and when should I prune it?"

Prune immediately after flowering by removing all the flowered stems. Tie in the new growths as soon as they are long enough.

"Are containerized roses worth the extra money?"

They will enable you to plant during the growing season, but need extra care while they establish.

"I intend to buy some new roses in the autumn. Is there anything I should be doing now?"

The best and most enjoyable job is to visit rose fields and public gardens to get an idea of what you want and how they perform.

"Can you suggest some rose varieties suitable for showing?"

'Alec's Red', 'Red Devil' and 'Sea Pearl' are all popular with exhibitors.

"What are patio roses?"

These are a type of short-growing, modern bush roses with a compact habit that makes them suitable for growing in beds near patios and paths and also in containers. Examples are 'Top Marks' and 'Sweet Dream'.

"Is it possible to grow miniature roses in window boxes?"

Yes. Choose small cultivars, such as 'Pour Toi' and 'Estru', and use soil-based compost.

"How do I deadhead miniature roses?"

Pinch out dead flowers with your finger and thumb. If all the flowers have faded, clip the bush lightly with shears.

"I bought some roses in pots in late winter. They had been chopped off at the top, and some of them are now dying back. Will they survive?"

If they were not pruned back to a bud, the piece of stem above the first bud will often die back. Remove this area immediately to prevent damage to the healthy shoots and leaves below.

"There are seed heads on many of my roses. Can I grow new ones from these?"

Yes, but they will be nothing like the parent. It will be years before you know what sort of plant you have produced, but it can be interesting to see the results.

"Can I take cuttings from roses?"

Many bush, shrub and climbing roses can be propagated this way. Take cuttings about 30 cm (12 in) long of flowered shoots in late summer and bury them up to half their length in good garden soil. You will be able to plant them out in their permanent positions late next autumn.

"I have a conifer hedge nearly 4 m (12 ft) high. How far can I cut it down without killing it?"

Conifers can be cut back as long as there is plenty of green growth left. If your hedge is not bare at the base, you can successfully remove 1.2–1.5 m (4–5 ft), as long as you shape the sides up afterwards.

"What would be a suitable architectural and slow-growing conifer for an alpine bed?"

Juniperus communis 'Compressa' is a slow-growing narrow conifer that coordinates well with both alpines and heathers.

"My *Lavatera* has died after only five years. Did I do something wrong when I cut it right back this spring?"

The shrubby mallow (*Lavatera × clementii*) is a short-lived shrub that often dies if it's pruned too hard. Shorten the branches a little in autumn to prevent wind rock, then prune back a little more in spring.

"I have a large evergreen *Ceanothus* and would like to prune it back. When and how should I do this?"

Evergreen shrubs flowering until midsummer should be pruned after flowering; those flowering later in the year are best pruned in spring. Thin out overcrowded branches and cut back the rest to promote new growth.

"I saw an interesting shrub with green stems and tassel-like flowers. What was this?"

This is the pheasant berry or Himalayan honeysuckle (*Leycesteria formosa*). It is a good shrub for attracting birds and produces self-set seedlings readily.

"When and how should I prune a *Clematis* 'Nelly Moser'?"

It should be thinned out after flowering, but usually these plants are in such a tangle that this is not possible. Leave it for this year then cut it back hard next spring; it will then flower later that summer.

"My clematis gets brown leaves all up the stems as the summer progresses. What is the matter with it?"

This is to be expected with most climbers, and no amount of pruning will ever fix the problem. Try planting a shrub in front of it if you feel it looks unsightly.

"Can I eat the fruit on my *Passiflora caerulea?*"

They are edible but not particularly tasty – although they do make good wine.

"My climbing hydrangea (*Hydrangea anomala* subsp. *petiolaris*), which has been in the ground for three years, refuses to climb. What can I do to encourage it?"

Leave it alone and do not be tempted to tie it to the wall. Eventually, some shoots will start to cling.

"When does a climbing hydrangea start to flower?"

It can take several years for this climber to establish itself. Usually once it starts to climb, it will also start to produce flowers.

"I am told I should cut my flowering almond back hard every year. When should I do this?"

Prunus triloba flowers on new wood produced after flowering in spring, so it should be cut back in early summer to keep it tidy.

"How do I keep a *Cotoneaster* × *watereri* 'Pendulus' tidy and a reasonable size?"

Remove all the new growth in late summer. You will also be able to see the berries better this way.

"My beauty bush (*Kolkwitzia amabilis*) never flowers. I prune it in spring to keep it from smothering other plants. How can I make it bloom?"

You are trimming off all the wood that will produce flowers. If it is not possible to leave it unpruned for a few years, it may be better to remove it altogether and start again with a more suitable shrub.

"Can I take cuttings from my corkscrew hazel?"

All corkscrew hazels (*Corylus avellana* 'Contorta') are grafted on to common hazel and will not strike from cuttings.

"Can I prune back an overgrown *Brachyglottis* bush?"

Yes. It can be cut back almost to ground level in early summer every year.

"My *Rubus cockburnianus* is a tangle of live and dead wood. How can I tidy it up?"

Remove all the old and dead branches in spring to encourage the production of new, white stems for winter decoration. You can still do this job in summer.

"I bought a lovely *Skimmia fortunei* last autumn, but it looks poorly now. What is wrong with it?"

Skimmias need neutral or acid, humus-rich soil and partial shade. They will not thrive in alkaline, chalky or dry conditions. Try growing it in a pot of ericaceous compost in a cool, shady spot.

"When should I clip a beech hedge?"

The best time for trimming beech (*Fagus*) is late summer or early autumn, so that it will not regrow before spring.

"I have an established lilac hedge. When can I prune it so it flowers next summer?"

Prune lilac (*Syringa*) immediately after flowering, but do not cut back hard.

"Can you recommend an easy dwarf hedge for a summer show?"

You cannot do better than the yellow-flowered *Potentilla* 'Jackman's Variety'.

"Can I use *Photinia × fraseri* 'Red Robin' as a hedge?"

It should give you a good evergreen hedge. Trim as the first flush of growth turns from green to red, and again in the second half of summer to maintain the red leaves.

"I love the look of a beech hedge, but my soil is poor. Is there an alternative?"

Plant a hedge of hornbeam (*Carpinus betulus*) instead. It will give you the same look, but it tolerates poor conditions better.

"I've heard of something called a 'fedge'. What is this?"

This is a fence that has been totally covered with a climber so it can be treated as a hedge. A good example is an overlap fence covered with ivy, which can be tidied with shears and looks exactly like a hedge.

"When should I plant dahlia tubers?"

Plant them in late spring to early summer, after the risk of frost.

"Can I take cuttings from dahlias?"

Yes. Start the tubers into growth in the greenhouse in spring. The shoots that appear can be rooted.

"My dahlia tubers are huge. Can I divide them?"

Yes. Dahlia tubers can be divided every other year to increase your stock.

"I divided my dahlia tubers last year, but many of them did not produce shoots. Why?"

Each piece of tuber must have a piece of the old stem attached.

"Should I start my dahlia tubers into growth before planting them?"

It will get them growing in the soil a little earlier, but it is not vital.

"How deep should I plant dahlia tubers?"

Plant taller cultivars 8 cm (3 in) deep, and bedding and shorter types 5 cm (2 in) deep.

"I have no luck keeping begonias after the first year. What should I be doing differently?"

At the end of summer, once all the foliage has dropped off, remove them from their containers and discard the old compost (which may contain the eggs of vine weevil grubs). Store them in a single layer in a cool, dry, frost-free, dark place in sand or peat substitute until spring.

"I think I've planted my begonias upside down. Will they grow?"

Not well. The slightly convex part of the tuber should be at the top; the rounded side is the bottom.

"How can I grow a pineapple lily (*Eucomis*)?"

Plant the bulb in a container and give it winter protection. Alternatively, grow it in a well-drained, warm, sunny border.

"I have no luck growing lilies in the garden. What conditions do they like?"

Most species of lily prefer neutral or slightly acid, humus-rich soil. Modern hybrids can cope with a wider range of conditions, but in all cases, the soil must not be overwet.

"I bought some lily bulbs that were being sold off cheaply. Is early summer too late to plant them?"

No. Plant them immediately in the border or in pots. This year only, they may flower later than normal.

"My soil does not suit lilies, but I love to see them appearing among other flowers in the herbaceous border. What can I do?"

Plant about six bulbs in a large pot. When the lilies are about to flower, put the pot in the border.

"How do I get a succession of gladioli flowers throughout the summer?"

Plant the corms at fortnightly intervals through spring to early summer.

"I have seen a purple, gladiolus-like flower in a park. What is this?"

This is *Gladiolus communis* subsp. *byzantinus*. It is fairly hardy and will usually survive temperatures below freezing without lifting.

"The leaves and flowers of my gladioli usually get silvery flecks all over them. What causes this?"

These are caused by thrips (thunderflies), which are particularly troublesome in a hot summer. Keep a watch for them and spray with an insecticide as soon as you see them.

"How can I stop my gladioli from flopping over?"

The only way to prevent this is to stake each plant at the time the corm is planted.

"I would love to grow freesias outdoors. Is this possible in a temperate climate?"

For reliable flowers, you will need to buy specially prepared corms. Plant them 5–8 cm (2–3 in) deep in late spring and early summer for late-summer flowers. They are tender plants, and you could grow them in containers, which should be kept in a frost-free place in winter.

"What is the tall, hyacinth-like plant that flowers in late summer?"

This is *Galtonia candicans*. You can plant the bulbs in spring or even early summer. They are hardy, and for best results you should leave the bulbs undisturbed once established.

"Can I leave a canna in the ground all the year round?"

No. Cannas are half-hardy and must be lifted and dried off in autumn before being stored in a frost-free place.

"I bought a lot of cannas in pots last year, but many of them look poorly this summer. What is wrong with them?"

They may be suffering from a virus. There is no cure and you should discard diseased plants. If you are buying cannas in full leaf, reject any that have stunted, streaked or deformed leaves.

"Why do my arum lilies (*Zantedeschia aethiopica*) never flower?"

The flowering shoots probably get damaged by frost. Grow them in pots in a cool greenhouse and put outdoors in late spring.

"Can I grow arum lilies by my garden pond?"

Yes, as long as the crowns are covered with about 15 cm (6 in) water, which will protect them against frost.

"I bought several eye-catching calla lilies in flower at a garden show two years ago. Why haven't they flowered since then?"

Like *Zantedeschia aethiopica*, they need humus-rich, permanently moist soil, but are much more difficult to get into flower after the first year.

"What do I do with a ginger lily (*Hedychium*) I bought this spring?"

This is generally used in the same way as the canna, as a bedding or 'dot' plant, and should be given winter protection.

"Can you suggest miniature summer bulbs for my rockery?"

Rhodohypoxis and *Bletilla striata* make lovely specimens for a sunny rock garden.

"Are there any summer bulbs I can still plant in early summer?"

Tigridia, chincherinchee (*Ornithogalum thyrsoides*) and *Childanthus fragrans* can still be planted at this time.

"What summer bulbs are there other than gladioli?"

Try *Sparaxis*, *Tritonia* ('Blazing Star') and *Ixia*. Plant them in autumn in full sun and well-drained soil.

"My grass is bare and yellow in places where it grows over what was a bricked area. What can I do to improve it?"

You will never get good grass if the soil is too shallow. To save time and money, take out the bricks, fill the hole with good topsoil and reseed or returf the area.

"In dry weather, is it necessary to use the sprinkler on the lawn?"

Not if the lawn is established. The grass will soon recover from drought once it rains.

"Is it possible to turf a lawn in summer?"

Yes, but it will need watering in dry spells, or the turf will shrink and cracks will appear.

"Could I seed a new lawn at any time in summer?"

You could have problems if the weather turns hot and dry, so it is best to wait until autumn.

"How can I get rid of the dandelions that keep appearing in my lawn?"

These should be spot-treated with a lawn weedkiller as soon as you see them or the leaf rosettes will smother the grass around them.

"What is lawn sand?"

This is a mixture of horticultural sand and ferrous sulphate. It is applied in late spring and early summer to control moss and young weeds and improve surface drainage.

"What's the best way to deal with clumps of coarse grass in my lawn?"

Slash clumps of weed grasses with a sharp, long-bladed knife or turfing iron before mowing, and they will eventually disappear.

"When can I start lowering the blades of my mower after the first spring cuts?"

Early summer is the best time. The blades should be lowered gradually.

"My lawn looks really tired. What fertilizer should I use?"

A quick green-up liquid feed is best. Use it monthly until midsummer, then apply an autumn fertilizer.

"What is the difference between quick-release, slow-release and autumn lawn feeds?"

A quick-release feed will get the grass growing quickly, but the effect wears off after a few weeks, and it needs to be applied again. A slow-release fertilizer will release nutrients slowly over a long period. An autumn feed contains less nitrogen and more nutrients to toughen the grass and feed the roots.

"When is the latest I can apply a summer lawn fertilizer?"

Midsummer is the latest you should do this.

"I applied a granular fertilizer with an applicator, and now I have darker green stripes in places."

You overlapped the application, so these areas received a double dose. The overfed grass will settle down in time.

"I put a weed-and-feed product on my lawn, but then the weather turned dry and I have a lot of brown patches. Will they regrow?"

Water immediately after application in future. After rain, most of the brown patches should recover.

"I seeded a new lawn in spring, but the weather turned dry and little grass has appeared. What shall I do?"

It should have been watered thoroughly during this time. Once you get a good shower, most of the seed should germinate. Keep it well watered for the rest of summer.

"When should I start feeding a new lawn?"

Do not use a powder or granular fertilizer for the first year, unless the instructions say you can. If growth slows, use a half-strength liquid feed every fortnight until the end of summer.

"My newly seeded lawn has more weeds in it than grass. Can I use a lawn weedkiller?"

No, you will probably kill the grass. If you prepared the site properly, most weeds will be annuals and will mow out, and with regular mowing, the grass will thicken up to fill any gaps.

"I intend to take a fortnight's holiday in midsummer. Can I leave the lawn uncut?"

Yes, but raise the blades again for the first cut after your return. Remember that an uncut lawn is a giveaway to burglars that the house is empty, so ask a friend to cut it once or twice if you are away for an extended period.

"Patches of couch grass are appearing in my new lawn. How can I get rid of them?"

Couch grass cannot cope with close, regular mowing and will soon mow out.

"My rotary mower regularly scalps parts of my lawn. What can I do to prevent this?"

Raise the cut for the time being and deal with bumps in the lawn in the autumn or next spring.

"I have used a lawn weedkiller on my grass. Can I compost the clippings?"

It is best to wait six weeks before composting treated grass mowings.

"Last year I put mowings that had weedkiller on them on my compost heap. Will the compost still be all right to use?"

Wait until next spring before using the compost, and keep it away from tomato plants, which are susceptible to even minute doses of weedkiller.

"Why do the tips of the grass in my lawn always turn brown after mowing?"

Either you are using a rotary mower and the blade is blunt, or the cylinder of your cylinder mower needs adjusting so that it cuts properly against the bottom plate.

"My lawn tends to get waterlogged at the lowest point after heavy rain. What can I do to improve drainage?"

Try spiking it with a hollow-tined fork or lawn spiker. If this does not work, you may need to construct one or more soakaways made of rubble and gravel under the topsoil to take away the surplus water.

"After cutting, why is my lawn covered in a series of ribs?"

Ribbing can be caused by the blades turning too slowly, or, more commonly, because the grass has grown too long between cuts. Walk more quickly when mowing, and mow more frequently.

"My lawn edges always look ragged. How can I improve their appearance?"

Use an aluminium or plastic edging, but make sure you sink it slightly below the lawn surface, or you will damage your mower blades.

"How often should I use an edging iron on my border edges?"

If you cut your edges regularly and repair any damage immediately, it should not be necessary to re-edge the lawn regularly.

"I like neat lawn edges, but my borders keep getting bigger."

Put away the edging iron. Returf the excess border in the autumn and use edging shears to keep the edges tidy in future.

"What do you use a besom for?"

Birch besoms used to be recommended instead of a yard broom for sweeping leaves and spreading worm casts. Nowadays, a flat-headed, plastic brush is more effective.

"I planted a new asparagus bed last year, and there are a good lot of spears coming through. Can I start cutting them now?"

You can take one or two, but the majority should be left for another year, or you will weaken the plants.

"Do I need a proper asparagus knife to cut the spears?"

No, any sharp, long-bladed knife will do.

"What height should asparagus spears be before I cut them?"

They need to be 10–12 cm (4–5 in) long.

"My asparagus spears appear a few at a time. How can I get enough for some decent servings?"

Cut each spear when it is ready and store in the salad drawer of your refrigerator until there are enough. They will last several days here.

"When should I stop cutting asparagus?"

Stop cutting no later than midsummer, so the plants have time to recover.

"Should I pinch out my runner beans when they reach the tops of their supports?"

Yes. If you do not, they will hang down over the shoots twined round the supports. However, they will still produce beans, so it will not affect the yield in the long run.

"Can I grow runner beans in a container?"

Yes. Use a deep container, such as a half-barrel, so that you can push the supporting canes in far enough. These should be tied together at the top to form a wigwam.

"Can you suggest a runner bean attractive enough to grow in the flower border?"

'Painted Lady' has flowers that are a mixture of red and white, and 'Hestia' is a dwarf cultivar with similar flowers.

"Many of my broad (fava) beans have a black mark where they were fastened to the pod. What is this?"

This is a sign that they are too old to be palatable, and the skins will be tough. Pick when the beans are young, bright green and sweet.

"Is it true you can eat young broad beans in their pods?"

Yes. They are delicious, almost like asparagus.

"My maincrop peas have finished and I want to sow some more as a follow-on. Can I use the same variety?"

No. They will mature too late to set a crop. Sow an early variety, such as 'Feltham First'.

"Why do my beetroot (beet) sowings always come up too thick?"

This is because each seed of many varieties is actually a cluster of seeds. Sow two seeds 10 cm (4 in) apart to make thinning easier, or grow a more modern variety, like 'Monopoly', which produces seeds singly.

"Why do I never get tight Brussels sprouts?"

'Blown' sprouts are usually caused by the plants whipping around in strong wind. Grow them in a sheltered place, plant deeply, and firm the plants as they grow. Support them with canes if they are grown in exposed situations.

"How can I stop spinach running to seed?"

Sow thinly, thin out the seedlings, keep moist and pick while young. Use the thinnings in salads.

"Why are my maincrop potatoes always ruined by blight?"

Try the new blight-resistant variety 'Sarpo Mira'.

"My onions always have thick necks and don't store well. What can I do?"

Use a high-potash fertilizer instead of a nitrogen-rich one. Sow the seed less deeply in future.

"How can I prevent scabby potatoes?"

Common scab does not affect the taste and cooking qualities of the crop. Grow a resistant variety, such as 'Wilja', and do not lime the soil.

"Can I grow watercress in my pond?"

Yes, but it can become contaminated if your pond is not scrupulously clean. Grow land cress instead, which is similar, but is grown in the vegetable garden.

"Why are my onions running to seed?"

This often happens if they get a check to growth, such as a long dry spell early in the season or a cold spell in late spring. Use these onions first because they will not keep.

"My Florence fennel has not bulked up at the base and is now producing flowerheads. What went wrong?"

You are growing them too close together. Thin the seedlings to 30 cm (12 in) apart and earth up the 'bulb' when it gets to the size of a table-tennis ball.

"My Jerusalem artichokes are getting tall and take the light from the crops near them. Can I shorten the stems?"

They can be shortened back a little, but if you cut them back too hard, the size of the tubers will be affected. Move them to a more suitable position next year.

"What causes root crops, like potatoes, carrots and beetroot (beet), to split?"

Splitting usually occurs when a substantial period of rain follows a drought. Try not to let your root crops dry out during dry spells.

"I planted some seakale in the spring. What do I do now?"

Water thoroughly and apply a weekly liquid feed throughout summer. Remove any flowerheads. Cut back the yellowing foliage in autumn and blanch in early winter by covering the dormant plant with a bucket. Pick the blanched shoots in spring.

"Can I grow a fig tree outdoors in a pot in my small garden?"

Grow *Ficus carica* 'Brown Turkey' in a big tub (a half-barrel is ideal) filled with soil-based compost. Stand the tub on bricks so the plant does not root through the drainage holes into the soil underneath.

"My new house has a paved garden, which looks bare. Can you suggest some year-round interest I can grow in pots?"

Use large pots of soil-based compost and plant a nucleus of variegated forms of evergreen shrubs, such as hebes, *Euonymus fortunei*, osmanthus and holly (*Ilex*). All these can be pruned regularly to keep them in check. Underplant with spring bulbs and fill out with bedding plants.

"Should I repot my pot-grown bay trees (*Laurus*) and box bushes (*Buxus*) annually?"

Once the plants have grown a reasonable size, replant them into large, sturdy containers filled with soil-based compost and they should not need repotting again.

"How do I make a standard bay?"

Choose a young plant with few side or basal shoots. Pot it on into a 30 cm (12 in) pot of soil-based compost and provide it with a cane. Cut back any side-shoots to about 5 cm (2 in) and remove all basal shoots, leaving the top to continue growing to the required height. Then remove the side-shoots and shape the head into a ball with secateurs if required.

"My cat sleeps in my window box and spoils it. How can I stop her?"

Cut pieces of holly and work them in between the plants.

"Should I feed my permanently planted tubs with a dry fertilizer as well as a regular liquid feed?"

They will benefit from a top-dressing of a slow-release, balanced fertilizer between early spring and early summer each year.

"Can I grow a rose in a container?"

Yes. Use a large container of soil-based compost and feed and water regularly. Bush roses generally succeed better than climbers.

"There is a concrete path all around my house. Can I grow a clematis against the house wall?"

Clematis in containers need good-quality, soil-based compost, which must never be allowed to get even slightly dry. Feed with rose fertilizer in spring and summer.

"I have nearly 100 pots and other containers, and watering them every day is getting too much for me. What can I do?"

Consider an automatic or semi-automatic watering system. There are many on the market that work well and are easy to fit.

"What can I do to conserve water?"

Never put water from sinks down the drain. Use a washing-up bowl and decant it into a watering can. Divert bath and hand-basin water into a butt if possible.

"Can I save the water from my washing machine and dishwasher to use on my tubs?"

No. The chemicals in this would be harmful if used regularly.

"Do hardy perennials work well in containers?"

They look good from spring to autumn, but there will be little to see in winter. Add one or two dwarf evergreen shrubs and spring bulbs to extend the season.

"How many plants can I put in a 35 cm (14 in) basket?"

Five or six are generally enough. This will allow for growth.

"My trailing petunias are always thin and straggly. What can I do to improve them?"

Pinch them out two or three times at the beginning of the season to encourage them to bush out.

"Why do my trailing petunias start to die off early in the season?"

This is usually because snails have been at work near the base. Put two or three slug pellets on top of the compost from time to time.

"How do I plant up a ball basket?"

Line and plant up the base of a deep wire basket, including the sides. Then line a similar basket and fill it with compost. Carefully turn it upside down, using a piece of card to stop the compost falling out. Place it on top of the other basket, remove the cardboard and plant the sides and base (now the top). Tie the basket together with wire or cable ties and attach chains somewhere near the top.

"Is it better to mix plants in a hanging basket or use the same kind?"

This is a matter of taste. Both schemes are effective.

"Do I always have to plant the sides of a basket?"

If you include suitable trailing plants, this is not always necessary. Solid decorative baskets are popular these days, and mean you do not need to plant the sides.

"I hear that it is environmentally unfriendly to use moss to line a basket. Why?"

Some moss is collected from environmentally sensitive areas which can upset the balance of nature.

"Apart from moss, what is the best liner for a hanging basket?"

Liners made from recycled cloth are effective and unobtrusive.

"Do slow-release fertilizers work in hanging baskets?"

Yes. They are designed to release nutrients as the temperature rises.

"Do I need to feed with a liquid fertilizer if there is already slow-release feed in the compost?"

A basket in full growth requires a lot of nutrients, so it does no harm to give it a liquid feed as well.

"What is the best soluble fertilizer for a hanging basket?"

Tomato food or a specific container feed is most effective.

"How often should I water a hanging basket?"

In normal summer weather, once a day may be enough, but in hot, sunny weather, they need watering two or three times a day.

"I find it a nuisance to keep my hanging baskets moist. Are there any ways of cutting down on watering?"

A basket with a reservoir will need watering less, but may become overwet in a rainy spell. Mix water-retaining granules into the compost.

"My baskets seem to look tired early in the season. How can I keep them looking good for longer?"

Never forget to feed and water, and deadhead regularly to give a succession of flowers.

"Would herbs thrive in a hanging basket?"

Yes. Herbs make ideal basket plants. Pinch or cut back regularly to keep them neat.

"Should I remove bubble insulation in summer?"

It will help to keep the greenhouse cool, but it will also reduce light levels, especially if hard-water deposits and algae have made it opaque, so it is best to remove it.

"I have just bought a greenhouse. How can I prevent it from getting too hot during the day?"

Whitewash the outside with greenhouse shading paint. Make sure all the vents are fully open during the day: invest in automatic openers. Leave the door open all summer.

"What is a sensible number of vegetables to grow in my 1.8 x 2.4 m (6 x 8 ft) glasshouse?"

Three cordon tomatoes, such as 'Shirley', two sweet peppers, one cucumber and a hot pepper should fit in and give you enough of each over summer without creating a glut.

"Can you recommend a good aubergine (eggplant) for my greenhouse?"

F1 'Moneymaker' is early, reliable and a heavy cropper. These are regularly sold as plants in the garden outlets in late spring and early summer for potting on at home.

"Can I grow courgettes (zucchini) in the greenhouse?"

Yes, but they take up a lot of room and they grow quite happily outdoors.

"What is meant by cordon tomatoes?"

These are plants grown on a single stem. Any sides-hoots are removed as soon as they are large enough to pinch out.

"When should I start feeding my tomatoes?"

Start regular feeding as soon as the first truss is setting.

"What is the best way to support cordon tomatoes in growing bags in the greenhouse?"

Tie one end of a piece of twine round the bag, close to each plant, and the other end to the roof framework of the greenhouse. Twist each plant loosely round the string as it grows.

"What is the best feed for a grapevine?"

Tomato feed will help to produce good crops of large grapes.

"Why are my greenhouse melons slow to ripen?"

Reduce the watering once they reach a good size.

"My tomato plants produce long flower trusses, but why do so few of the flowers set and produce fruits?"

This is usually because of poor pollination. Tap the supports or mist the plants at midday.

"I have a peach in a pot in the greenhouse. It flowers profusely, but why do most of the fruits drop off before ripening?"

You may be forgetting to water regularly. Otherwise, it possibly needs repotting into a bigger container.

"Are there any plants I can grow during summer in my south-facing conservatory?"

If you shade the roof with blinds or voile and keep all the windows open, most conservatory plants will grow quite happily if you feed and water regularly and group plants together so they produce their own microclimate.

"Why are my conservatory-grown spider plants (Chlorophytum) turning brown at the ends?"

This is a sign of stress, caused by being potbound, not being watered frequently enough or being placed in a hot situation, such as high up in the conservatory. You should try to remedy, if possible. If you find the brown ends unsightly, cut them off.

"The dragon plant (*Dracaena*) in my conservatory is getting rather large. Can I cut it back?"

Yes. Cut to a bump on the trunk and it should shoot from there, and possibly lower down as well.

"Can I take a cutting from my dragon plant?"

Some of the younger leaf rosettes will root in a mixture of houseplant compost and sharp sand. You can also try cutting the bare stems into pieces and pushing them into a similar compost in a heated propagator with the top off.

"What is the best way to train jasmine and bougainvillea?"

After pruning, train the new shoots in a loop around a wire framework or trellis.

"All the leaves started to fall off my bougainvillea, and I found the pot was standing in water. Will it recover?"

More plants are killed by overwatering than underwatering. Allow the compost to dry out and the bougainvillea will usually recover.

"How often should I water my conservatory plants in summer?"

When they need it! If the compost feels dry and the pots are getting lighter, they need watering.

"My *Phalaenopsis* orchids have developed bleached patches on their leaves. What is the cause?"

They have been in too much sun. Move them to a shadier part of the conservatory in summer.

"Could you recommend a really fragrant, summer-flowering plant for my conservatory?"

You can do no better than heliotrope. Although often sold as a hanging basket plant and discarded at the end of summer, in a conservatory with some winter heat, it is rarely out of flower.

"Why have several of the cacti in my greenhouse gone rotten at the base?"

You are overwatering. Do not stand the pots in water and let the compost dry out completely between waterings.

"Which vegetables would you recommend my two children, aged seven and nine, try growing to start off with?"

Try quick-growing plants, like 'cut-and-come-again' lettuce, baby carrots, radishes and mixed salad leaves.

"Can you suggest a simple project to get my three-year-old daughter interested in growing things?"

Cut the top off a carrot, place it in a little water in a saucer and watch the ferny leaves sprout.

"What indoor plants might interest children?"

Living stones (*Lithopes*) are a type of succulent that many children like to grow and even collect.

"I have a rough piece of ground at the bottom of the garden. Would it be suitable for my children to have their own garden there?"

This is not a good idea. If plants do not grow well for them, it may spoil their interest in gardening. Give them a small, sunny patch near the house, or a small raised bed on the patio.

"Could you suggest some pretty flowers my children could grow from seed?"

All hardy annuals grow quickly and easily in a sunny spot. You may need to supervise when they are sowing and thinning out.

"What plants might my children be able to grow from pips and stones?"

Try apples, pears, peaches, avocados, grapes, dates and lychees. All these can be germinated in multi-purpose compost.

"My children want to grow something really useful indoors. What do you suggest?"

Buy them some packets of edible sprouting seeds. Start with a glass jar and some muslin to grow them in. If they remain interested in a month or two, buy them a proper seed sprouter.

"My young son is keen on gardening and has just bought his own mini-greenhouse with his pocket money. What should he start off with?"

Showy, edible crops, such as red and yellow sweet peppers and tomatoes, should get him going. One plant of each is enough to start with.

"My children want to see what happens when they plant a peanut."

This is a fascinating plant, producing a shoot rather like a pea. After flowering, it bends over and buries itself into the compost, and then the peanut pod forms below the surface.

"My grandchildren have grown little oak trees from acorns and want to know what to do next."

Help them to pot them on until they have grown to about 60 cm (24 in). Then suggest they offer them to someone with enough space to accommodate them, such as a farmer with a woodland planting.

"My children want to plant a conker. Will it grow?"

Plant conkers, the nuts of the horse chestnut tree (*Aesculus hippocastanum*), in small pots of compost in autumn. Young trees will emerge next spring. Unfortunately, the trees eventually get too big for the average garden, but can be planted in the wild.

"Can you suggest something I can amuse my three grandchildren with when they stay with me for their summer holidays?"

Give them a tomato plant each and have a competition to see who grows the most fruit by weight. Make the prize something horticultural, and the winner could be hooked on gardening for life.

"My small son wants to grow the tallest sunflower. What should he do?"

Buy seed of a giant cultivar and sow it in a small pot of compost in late spring to early summer. Pot it on until you have a large specimen ready for planting in the garden. Plant it by a sunny wall and support it as it grows.

"What is the best sort of pond liner?"

A flexible liner is easier to fit than a rigid one. Butyl is expensive, but lasts for 30 years or more and is easy to fold to the shape of the pond.

"I like the sound of water in the garden, but I am concerned for the safety of my young children. Is there an alternative to a pond?"

A simple water feature, like a fountain trickling over cobbles, will give you what you need until the children get older, at which point you can decide if you want to convert it into a garden pond.

"Do I need a filter in my pond?"

If you intend to have fish you will need a filter to keep the water healthy. Otherwise it is not necessary.

"I have had a water lily in my pond for several years, and although it produces leaves, it never flowers. Shall I start again?"

Water lilies often do not flower if they are planted too deep to start with. Plant in a crate and stand it on bricks in the pond until the crown is about at water level. Lower slowly as the leaves develop until it is at the proper depth.

"What sort of compost should I use for aquatic plants?"

A heavy loam is best. If this is not available, most water garden specialists sell bags of aquatic compost.

"Is it better to use planting crates or to put a layer of soil in the bottom of the pond?"

You can control the plants better in crates. They also make it easier to clean the pond out.

"When I sink my crates in the pond, a lot of fine soil washes out."

Line the crates with hessian (burlap) and cover the surface with washed pea shingle.

"My aquatic supplier sells marginal plants in plant pots. Can I leave them in these pots when I put them in my pond?"

No. The soil will turn foul and smelly, and the roots will rot. Replant them in proper planting crates.

"I have a tiny pond. Is there a water lily small enough for it?"

Try *Nymphaea tetragona*, which is small enough to grow in a sink.

"I think my water lily is too big for my pool. The surface is covered with leaves, which are pushing each other up. What should I do?"

Remove it and replace with a more moderate grower, such as *Nymphaea* 'Laydekeri Alba'.

"I put some ivy-leaved duckweed in my pond, and it has taken over everything. How can I get rid of it?"

Duckweed will soon cover the surface but it doe keep the water clear and is a useful food for fish. Clear it regularly from your pond; duckweed is good for the compost heap.

"What should I use to waterproof a leaky concrete pond?"

Fill in any cracks with mortar, then apply at least two coats of a proprietary pond waterproofing compound. Leave for a few days before refilling.

"What precautions should I take before introducing fish to my new pond?"

Fill it with water and empty it at least once to remove any chemicals that might be harmful. When you are using tap water, leave it a few days for the chlorine to evaporate before adding the fish.

"Should I add the plants and fish at the same time?"

No. Allow a couple of weeks for the plants to settle down before any fish are added.

"What is the optimum depth for a garden pool?"

To keep fish in it over winter, a pool should be 45–60 cm (18–24 in) deep. It is not necessary to make it too deep because most water lilies cannot cope with deep water.

"I filled my new pool and the water turned green."

This usually happens after the first filling. Once a balance is achieved between water, plants and wildlife, it will usually clear, especially when the lily leaves shade some of the surface.

"How do I get rid of blanket weed?"

Moving water, for example a pond with a fountain, tends to be less troubled with blanket weed, which usually attaches itself to the side and can be pulled off by hand. Blanket weed is best raked off regularly.

"Are there any chemicals I can add to my pool to control algae?"

Yes, but when you kill off the algae, the dead material can pollute the water. This is a particular issue when fish are present.

"How does a wildlife pond differ from a normal garden pool?"

A wildlife pond should be capable of attracting indigenous fauna – not only larger species, such as frogs, toads and newts, but also the larvae of the insects on which they feed, such as damselflies and dragonflies. For this reason, you should avoid steep sides and ornamental fish, which will eat much of the wildlife you are hoping to encourage.

"I never get frog spawn and tadpoles in my pond, although I see frogs in the garden."

Fish will eat much of the frog spawn. Otherwise, if the water is disturbed (by a fountain or similar) frogs and toads will go elsewhere. Turn off the fountain during the breeding season.

"How can I make my pond user-friendly to wildlife?"

Make a ramp into the water, so that wildlife can easily drink.

"How often should I clean out my pond?"

Every two or three years is usually recommended, but it depends on how clean the water is and how happy the plants and fish are. It may be only necessary to pump out the gunge at the bottom, allow things to settle and then top up with clean water.

"Should I clean out my wildlife pond?"

No, not unless something is seriously wrong. The more you upset the natural balance, the less wildlife will be attracted.

"My flag irises have punctured my liner. Can it be repaired?"

Strongly growing pond plants, like flag irises (*Iris pseudacorus*) and reed mace (*Typha latifolia*), have no place in a typical garden pond. Start again with a new liner and replace the flag irises with a more suitable species, like *I. laevigata*.

"I bought a collection of plants and fish for my pond, and included with this were some water snails. These are now eating my water plants. How can I get rid of them?"

Water snails keep the pool clean and are rapid breeders. They can cause problems with pond plants, but fish will control them by eating the eggs.

"What can I do to prevent lily beetles eating my water lilies?"

Wash them off with a strong jet of water from the hose and the fish will eat them. The same applies to water lily aphids.

"I would like a tiny pond in my paved garden. What can I do?"

Make a pool in a large half-barrel, lined with heavy-duty polythene. You can grow a miniature water lily and one or two marginal plants, like water forget-me-not and mimulus. Stand the marginal crates on bricks to raise them in the water.

"Can I keep goldfish in a miniature pool?"

Make sure that there is enough oxygen in the water in summer by planting one or two submerged oxygenating plants. They will need winter protection because the water may freeze solid.

"Can you suggest an easy way to deal with a redundant fish pond?"

Make some holes in the bottom, fill it with good topsoil and turn it into a bog garden. Suitable bog plants are sold in the aquatic section of most garden stores.

"How can I stop a heron from taking all my fish?"

Netting is the only effective solution, although it does spoil the appearance of the pond.

"How can I get rid of slugs and snails?"

Slug pellets, applied strictly according to manufacturer's instructions, are still the most effective way and are safe if properly used and stored. There are many organic controls, usually based on aluminium sulphate or copper, which have varying results but do have some effect. Biological controls are available online and from some garden stores; they are effective against slugs but not snails, and are expensive. Slug traps filled with beer can be used for localized control.

"Molehills are ruining my lawn. I have tried putting a plastic bottle on a stick but it does not work. What else can I try?"

There are many devices said to deter moles, some expensive and not always effective. Children's plastic windmills pushed into the runs seem to work as well as anything, but they need moving regularly because the moles get used to them.

"There is die-back in my privet hedge. If it is honey fungus, is the hedge doomed?"

Probably. Honey fungus (*Armillaria*) is a serious fungal disease of plants, that usually, but not always, affects shrubby plants. No cure is available to the amateur gardener. Some plants, including privet, are more susceptible than others. It is characterized by honey-toned toadstools around the base, and black, bootlace-like rhizomorphs in the soil. If you see these, remove the hedge and replace it with a fence. Do not replant shrubs or trees in the same soil, and don't plant near by.

"Why do my honeysuckle flowers die before they open?"

This is probably caused by mildew. Spray with a fungicide before the problem occurs, and be sure to keep the roots cool and moist.

"What has made the shoots of my honeysuckle distorted and sticky?"

Aphids are sucking the sap of the young growths. Spray with an insecticide immediately or flowering will be affected.

"Why is my Leyland cypress hedge turning brown all over?"

This can be due to conifer aphid, but is also a sign of stress in hard-clipped hedges. Regular spraying with an insecticide will control any bugs.

"What is causing the silvery channels in the leaves of my holly?"

This is caused by the holly leaf miner. It rarely causes serious problems, so treatment is not usually necessary.

"Some of my rose leaves have been skeletonized. What has caused this?"

The rose slugworm has caused this damage. It can be curbed by an insecticide, but will not kill the plants, so is best left alone.

"Why are my cherry leaves covered in brown spots and little holes?"

This is shot hole disease, a fungus that attacks members of the plum (*Prunus*) family growing in poor conditions. Improve cultivation if possible.

"Why are parts of my golden privet turning green?"

All-green branches often appear in variegated shrubs and trees. Remove them completely as soon as you see them, or they will take over.

"I have just spilled petrol on my lawn. What can I do?"

This will kill the grass. You will have to remove the turf and the contaminated soil underneath it, then replace it with clean soil and reseed. Never refill the mower on the lawn.

"What is damaging the stalks of my raspberries just below the flowerhead?"

Blossom weevils will partially sever stalks of strawberries and raspberries in this way before laying their eggs in the buds. Pick off and destroy the buds.

"What are the nasty, worm-like creatures in my potatoes?"

They are wireworms, which can ruin a potato crop. Dig the soil where the potatoes are to be planted several times during winter to expose these pests to birds.

"One of my apple trees has patches of deformed bark, which peel off to reveal bare wood. What is wrong?"

Apple canker is a serious disease. Cut off the affected branches to clean wood.

"Why are my peas covered in white, powdery patches?"

This is powdery mildew. Spray with a fungicide suitable for edible crops as soon as you see it. Badly affected plants should be taken up and disposed of (not composted).

"What can I do to prevent my peas becoming maggoty?"

These are the larvae of the pea moth. Avoid growing mid-season crops, which are usually the worse affected.

"The leaves of my *Daphne mezereum* are mottled and fall early at the end of summer. Why is this?"

Cucumber mosaic virus affects many garden plants, including daphnes. It will not affect flowering, but the plant may not live long.

"My outdoor tomatoes always start to rot toward the end of summer. What causes this?"

Potato blight also affects tomatoes. You will avoid this if you grow them in a greenhouse or conservatory.

"What causes many of my rose shoots to produce stems without flower buds?"

A late frost or poor nutrition are the usual causes of rose blindness. Always feed and water well. Cut back to a leaf with five leaflets.

"How can I stop earwigs spoiling my dahlias?"

The old-fashioned remedy of filling plant pots with hay or straw and supporting them, upside down, on canes among the plants, works as well as anything. Replace the straw every morning and burn it to destroy these pests.

"Many of my apples have brown, ribbon-like scars on the peel. What is wrong with them?"

Control apple sawfly by destroying all affected fruit. The grubs drop from the fruit to the ground in summer, so keep the soil clean and hoe frequently to expose the pupae to birds.

"Will cuckoo spit damage my plants?"

This frothy substance is made by the frog hopper, a sap-sucking insect that uses it to protect itself from predators. The insect does no serious damage; just wash the froth off if you do not like it.

"My pear tree started to die overnight and looks as if it has been burned. What is wrong?"

This sounds like fireblight, a devastating disease that can affect any member of the rose family. Cut off and burn affected branches immediately. Remove badly affected specimens completely.

"Ants are eating my roses. How can I stop them?"

Actually, they are not. They are 'farming' the greenfly for their honeydew. Control the aphids and the ants will go away.

"My beech hedge is covered with aphids, which makes trimming unpleasant. When is the best time to apply insecticide?"

Spray about two weeks before you want to trim the hedge.

"Why are my bay trees and camellias covered in soot? We do not have a fire."

This is a black mould (mold) that grows on honeydew (the secretions of aphids). The only way to remove it is by washing the leaves with soapy water. Spray aphids with an insecticide.

"My parsley has turned a reddish-brown and stopped growing. What is the cause of this?"

Root aphids affect many crops, including lettuce and members of the carrot family, like parsley. Do not grow similar crops in the same soil as affected plants.

"What is making the leaves on my bay tree sticky?"

This is a secretion from the bay sucker. Spray with an insecticide permitted for edible crops.

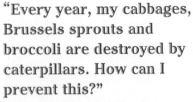

"Every year, my cabbages, Brussels sprouts and broccoli are destroyed by caterpillars. How can I prevent this?"

The caterpillars of the cabbage white butterfly can ruin a crop. Cover the crops with fleece as soon as you spot the first adult butterfly, and keep them covered for the rest of the season.

"What is causing the little holes in the leaves of my turnips?"

Flea beetles. Spray with an insecticide or cover with fleece for the life of the crop.

"Many of my cabbages have died. When I pulled them up, the roots were eaten and full of maggots. What are they?"

These are the larvae of the cabbage root fly. Surround the young plants with discs made of roofing felt or heavy-duty polythene so that the adult flies cannot lay their eggs near the plants.

"Why are my leeks covered with orange spots?"

This is leek rust. Pull up and destroy all affected plants, and grow your onions and leeks somewhere else next year.

"Why are all my greenhouse tomatoes developing black spots on the bottom of the fruits?"

This is blossom end rot, caused by a lack of calcium in the compost. Erratic watering makes the problem worse. Always water thoroughly and regularly, and make sure your tomato feed contains calcium.

"Why do my tomatoes never turn red at the top?"

This is a condition known as greenback, caused by too much direct sunlight. It usually occurs in older varieties, like 'Alicante' and 'Moneymaker'. Grow an F1 hybrid and shade the greenhouse.

"How can I stop my greenhouse vine getting covered with mildew every year?"

Increase the ventilation as much as possible and spray with a fungicide suitable for edible crops every fortnight from late spring.

"What is the best way to control greenhouse whitefly without chemicals?"

Hang yellow sticky traps among the plants.

"How can I control red spider mite on my greenhouse plants?"

Increase the humidity by damping down the floor on sunny days. Mist the plants regularly.

autumn

"Are there any green manure crops I can sow in autumn?"

Field beans can be sown in autumn. When they are dug in next year, they will fix nitrogen in the soil, meaning you need to add less high-nitrogen fertilizer. You can also sow Hungarian rye and tares. Look on the internet for suppliers.

"I use the no-dig method of cultivation in my vegetable plot. Can I still use green manures?"

Yes. Cut down the tops at the appropriate time and plant through them. The worms will incorporate the manure into the soil.

"I sowed some mustard as a green manure in the summer. What do I do now?"

Mustard should be allowed to grow for one to two months before digging in. Wait about a month before replanting, or allow the soil to rest over winter and plant in spring.

"I want to dig up part of my lawn and make a flowerbed. What should I do with the turf?"

The easiest way to deal with it is to bury it, upside down, about 23 cm (9 in) deep. The grass will rot to increase soil fertility at a lower level.

"How can I test the pH (acidity or alkalinity) of my soil?"

A soil-testing kit will give you reasonably accurate readings, but for really precise results, you need to send samples to a soil-testing laboratory.

"The soil in parts of my vegetable garden has become compacted. What is the best way of dealing with this?"

Divide the area in half. Take out a trench of soil, one spade deep, at the beginning. Fork over the base. Dig the next trench, putting the soil in the first one. Continue to the end of the area, then turn around and work back to the beginning, filling in the final trench with the soil from the first.

"How can I avoid compacting my soil when I'm digging in autumn?"

Don't work when the soil is wet, and try not to walk in the same place every time. If you need to plant while the soil is wet, stand on a plank to spread your weight.

"My grandfather didn't use concentrated fertilizers on his soil, and he had wonderful flowers and vegetables. Why do I need them?"

It is possible to keep the soil in good condition by using animal manure or garden compost, but you need a lot of it: about 50 litres to 5 sq m (11 gallons per 50 sq ft) of farmyard manure or twice as much garden compost. Although this is traditionally applied before digging in autumn, it is better added in the spring so the nutrients are not wasted.

"A friend of mine uses sawdust as a soil improver. Is this a good thing?"

Some sawdust can be added to the soil as an improver, but too much will upset the soil's texture and rob it of nitrates as it rots.

"I want to kill weeds with glyphosate before digging a new piece of ground this autumn. Is it safe to do this?"

Yes. Glyphosate is neutralized when it is in contact with soil. Just make sure all the weeds are completely dead before starting to dig.

"I want to get mechanized with my vegetable gardening. Is rotavating as good as digging?"

Not really. It is unlikely to work the soil as deep as digging, and it can cause compaction if used when the soil is wet. Furthermore, it will not leave the ground rough enough for frost to penetrate properly to improve the structure over winter.

"I have acid soil. How much lime should I add to make it less sour?"

Depending on how light or heavy your soil is, you will need between 500 g per sq m (1 lb per sq yd) and 1 kg per sq m (2 lb per sq yd).

"Does lime do anything other than alter the pH of the soil?"

The calcium it contains is a plant food. It also releases other plant foods from organic matter and can discourage pests and diseases.

"Which pests and diseases can be managed by adding lime to the soil?"

Slugs, wireworms and leatherjackets do not like lime. Clubroot disease can be curbed by periodic liming.

"Should I lime flower borders as well as the vegetable plot?"

It depends on what you want to grow and how acid the soil is. Most ornamental plants prefer neutral soil, but wallflowers (*Erysimum*), pinks and carnations (*Dianthus*), delphiniums and clematis do best in alkaline conditions.

"When I have applied lime once, why do I have to do it again from time to time?"

Rain gradually washes lime out of the soil, so it becomes more sour. In addition, adding large quantities of bulky manure regularly will increase the rate at which soil becomes acid.

"When should I apply fish meal to my vegetable garden?"

You can add it in autumn, before planting out overwintering crops, or in spring.

"Last year, I dug stable manure into the garden, and I got a crop of oats this spring. Why did this happen?"

Many grass seeds will pass straight through a horse's digestive system unharmed and ready to germinate. For this reason, it is best to add manure to the compost along with other garden compost materials.

"I recently removed a conifer hedge and shredded it. Can I dig the shreddings into my garden this autumn before planting new perennials?"

The shreddings will contain a lot of resin, which can be harmful in large doses. It is best to add it to the compost heap in small amounts or to use it as a mulch on mature shrubs.

"I want to replant a border that has been mulched with a thick layer of coarse bark chips. Can I dig this in?"

It is better not to. The bark will rob the soil of nitrogen as it breaks down, and make the soil so coarse that young roots will find it difficult to establish properly. Rake it off and reuse it.

"Is peat substitute a good soil conditioner?"

Peat substitute can be made from a variety of materials, most of which contain a lot of cellulose but few substances to stimulate bacterial activity. Added in moderation, it can improve soil texture, but too much may affect fertility.

"Why do many plants die in wet soil in autumn and winter?"

The water replaces air in the soil, so, effectively, the plant roots suffocate.

"I have a shallow soil over rock. Will increasing the depth by adding more topsoil help cultivation?"

Yes, as long as the original layer is well drained and the topsoil comes from a reliable source.

"My soil is full of perennial weed roots. If I roughly dig the ground in autumn, will frost in winter kill them?"

Some may die, but many more will be alive and well beneath the surface. It is best to dig in early autumn, let the weeds grow until late spring, then spray with glyphosate. There are no satisfactory short cuts to weed-free soil.

"How can I get rid of a wilderness of brambles?"

Hire a brush cutter and cut them to the ground. Wait until about 60 cm (24 in) of new growth has appeared, then spray with a brushwood killer or weedkiller for stubborn weeds.

"Is there anything I can do to combat a naturally high water table in winter?"

The easiest way is to complete all autumn digging early in the season, plant in spring when the table has started to go down, and choose plants that either have a shallow root system or that do not mind standing with their feet in water during the coldest months of the year. A good local nursery should be able to recommend suitable species.

"What is the difference between garden compost and leaf mould (mold)?"

Leaf mould is the substance produced by fungi breaking down leaves in autumn. Garden compost consists of many different plant and other organic materials broken down by the action of bacteria.

"What can I use leaf mould (mold) for?"

Leaf mould can be used instead of peat in acid composts and as a soil conditioner. It can also be added to heavy garden soil as a planting mixture for new trees and shrubs.

"What are the best leaves for making leaf mould (mold)?"

Beech (*Fagus*) and oak (*Quercus*) are said to make the best leaf mould, but, in practice, any tree or shrub leaves will do.

"Do I need to do anything with leaf mould (mold) before using it?"

If you are adding it to planting composts, put it through a riddle (coarse sieve) first to break it down finely.

"Do I have to compost tree leaves separately?"

No. They can be added to the compost heap, but they may slow down the rotting process unless they are mixed with green waste.

"Do I make leaf mould (mold) in a bin in the same way as compost?"

The fungi involved need air to break down leaves, so a wire mesh container is a better option.

"How old does leaf mould (mold) need to be before I can use it?"

It depends on the leaves. Some leaves will have formed leaf mould in 12 months, but with some hardwoods, like oak, it may take another year before it is ready.

"I want to use my compost, but the top layer has not rotted properly. What should I do?"

Remove the top layer and use it to start off the compost in your empty bin.

"What can I use garden compost for in autumn?"

Add it to the soil when you're digging. If the compost is coarse, put it at the bottom of a trench and cover it with soil. The worms will do the rest.

"Can I still add old newspapers to the compost bin from autumn to spring?"

Yes, but shred them or add them in small amounts, or they might prevent the compost from rotting.

"I am making a rose bed. Is it better to mix cultivars or stick to the same kind?"

A hotchpotch of cultivars, especially if they have different habits, can look rather bitty. Choose one cultivar only with similar heights and habits.

"What is the best thing to use to tie climbing roses to their supports?"

Use black plastic cable ties, available from electrical wholesalers and DIY shops. These are unobtrusive and can be cut if necessary when you need to release a branch.

"How long do I have to continue spraying roses against disease?"

It depends on the weather. If the roses stop growing and flowering in autumn, you can stop then. However, in some mild autumns and winters, the roses never stop growing until pruned, in which case, a three- or four-weekly spray may still be necessary.

"Should I put rose fertilizer on my rose bed in autumn?"

Most rose feeds are quick-acting and will either keep the roses growing when they should be dormant or will be completely wasted. If you want to feed your roses, use coarse bone meal or granulated (not powdered) hoof and horn fertilizer.

"My roses were badly affected by black spot this year and most of the leaves have now fallen off. Can I compost these?"

No. It is better not to because disease spores may be spread when you use the compost.

"I was tidying my rambler roses and found lots of new, long shoots that it seems a shame to cut off. What should I do with them?"

See if you can replan your garden so that these shoots can be trained on chains or wires alongside the lawn or over a flowerbed.

"My roses are flowering late this year. Will it spoil them if I do not cut them off?"

No. When you prune them, new flowering shoots will be produced.

"Can you suggest a rose for winter interest? Most roses look so drab."

Rosa sericea f. *pteracantha* has large, translucent red thorns. Autumn is a good time to plant it, as it will have a chance to make a good bush for the following autumn and winter.

"When a rose bush is planted, does the part of the plant where the branches come from go above or beneath the soil?"

This is where the rose cultivar was budded on to the rootstock, and it should be buried about 2.5 cm (1 in) below soil level. This is so the cultivar itself also makes roots, so when the bush is established, it will have a large, strong root system that will produce a healthy plant with many branches coming from near the base.

"I have just bought some bare-root rose bushes and nearly all the roots have been chopped off. Will the rose survive?"

This may have been for the convenience of the packer, but hard root pruning actually does no harm at all, as it encourages fibrous feeding roots to form.

"I bought some roses in containers for autumn planting. Why did the compost fall off as I was putting them in the holes?"

They may have been recently lifted and potted up. It does not matter at this time of year if the compost falls off; if you were planting bare-root bushes, they would have no compost on them anyway.

"I bought some bare-root roses at a supermarket. The roots were wrapped in hessian (burlap) and elastic bands, and the instructions said to leave these on. Should I?"

At this time of year, it is better not to. The roots will have been restricted by the wrapping and elastic bands, and will never make a good root system unless they are released.

"My discount store often sells prepacked roses in autumn that look like a good buy. Should I be tempted?"

Sometimes you do get good results with discount roses, but in general you are better off buying from a reliable garden store, or, preferably, a nursery specializing in raising roses. Do not buy them if shoots are starting to grow in the package.

"Can you suggest a rose for a silver wedding anniversary?"

The cultivars 'Silver Anniversary' or 'Silver Wedding' would be appropriate.

"What would be a good rose to give to a friend who has just had twins?"

The creamy-white and pink 'Double Delight' has a gorgeous scent and should fit the bill.

"Is there a suitable rose to celebrate my daughter's wedding?"

The rambler 'Wedding Day' would be most appropriate. Grow it on a pergola or fence and prune hard after flowering.

"Should I autumn-feed my lawn every year?"

Not if it is healthy. Every two or three years should be enough.

"When should I start raising my mower blades?"

Do this in early autumn.

"My lawn always starts to grow more strongly at the end of summer. Can I still cut it short?"

No. You should raise the blades to about 4 cm (1½ in) at this time of year.

"When can I start sowing my new lawn?"

Start at the end of summer, when there is more likelihood of rain and night-time humidity is higher.

"What happens if I sow a lawn late in the season and the seed does not germinate?"

Some of the seed will germinate in warmer spells in early winter, some seed will lie dormant until the spring, and some will rot.

"Should I water a new lawn in a dry season?"

Wait until it shows signs of stress. If this is not evident, it does not need watering.

"How late can I sow a new lawn?"

With warmer weather extending later in the season, it is possible to sow up to mid- to late autumn.

"I am told my lawn needs scarifying. What is this?"

Scarifying is giving the lawn a good raking to remove thatch (dead material) that has accumulated near the soil during the summer. This can be done with a spring-tine rake or with a mechanical (usually electric) grass rake.

"Why do I have to rake the grass?"

Thatch in the lawn can make it almost waterproof, preventing rain from getting to the roots, and so choking the grass.

"Why do you have to spike a lawn?"

Spiking allows rainwater and air to reach the grass roots more easily. If you have sandy soil or a good loam soil, it may not be necessary, but lawns on heavy soils benefit from spiking at least once a year.

"Can I use an ordinary garden fork to spike the lawn?"

Yes, but a hollow-tine fork, which takes out cores of soil, is better. Power-driven spikers are also available.

"I saw some spiking sandals advertised recently. It claimed you could spike your lawn and get good exercise at the same time. Is this true?"

Yes, but these are really only practical for a small lawn, because it is a slow and tiring job.

"What should I do with the soil that comes up when I spike the lawn?"

Remove it completely. It is best not to spread it in flower borders and vegetable beds because it may contain grass or weed seeds. Add it to the compost heap in think layers.

"Can I compost all the old grass and moss that is produced after raking the lawn?"

Yes, but make sure it is well mixed with other compostable materials, or it may slow down the rotting process.

"I get a lot of leaves settling on my lawn. If I cut them up when mowing, will they fertilize the grass?"

No. The debris will attract worms and may contribute to a build-up of thatch. Worms also attract moles, so always clear up leaves.

"I sowed a new lawn in early autumn. It germinated quickly and is now quite long. Should I mow it before the winter?"

It is better to leave mowing until spring, although if the grass is long and the weather forecast is for mild, dry weather, you can take just the tips off to tidy it up. Never do this after mid-autumn.

"I missed sowing my lawn in early autumn. Can I turf it later in the season?"

Yes. Mid- to late autumn is a good time to turf a lawn, but it will cost much more than seed.

"They are building on a local school's playing fields and the builder is offering the turf at a really good price. Is it a bargain?"

It depends on how the turf has been lifted and what part of the playing field it has come from. Avoid anything from pitch outfields, which will be much coarser than lawn turf and will need cutting several times a week to keep neat.

"My garden suppliers sell cultivated turf that is very thin. How can it possibly grow properly once laid?"

The thinner the turf is cut, the better it re-establishes. Thick turfs take much longer to root.

"Is there a cut-off date for turfing?"

No. It is possible to turf throughout winter if the weather is dry and mild, but turfing in wet or frosty weather should be avoided.

"Should I roll a newly turfed lawn?"

No. If it was properly laid on a bed of fine soil, it will knit without any pressure.

"My grandfather always gave his lawn a good rolling with a heavy roller in autumn. Should I be doing this?"

No. You will compact the soil and may cause growing problems and moss. The roller on the mower is quite adequate for rolling, but it is really not necessary to roll a domestic lawn at all.

"My lawn has become mossy. Should I apply lawn sand now instead of in the spring?"

No. It will make the grass grow too much, and it will then be prone to winter lawn diseases. If the grass is still growing, raise the mower blades higher and the grass will smother much of the moss.

"I remember buying a product that killed worms in lawns. Can you still obtain this?"

Thankfully, no. Worms are worst in lawns where the clippings are left on all the time and autumn leaves are not cleared up frequently. If you collect the clippings and rake up leaves daily, the problem will soon be reduced.

"How do I keep leaves out of the pond in autumn?"

Stretch a fine mesh over the surface until the trees are bare.

"I have been away for a while and my pond is full of leaves."

Remove those floating on the surface immediately. Then sift through the water with a pond net and remove any that have sunk and started to rot.

"I floated barley straw on my pond in the summer to control blanket weed. Do I leave it in over winter?"

Barley straw pads are said to control algae and blanket weed, although not everyone finds them effective. It is best to remove them in autumn so they do not rot and pollute the water.

"Should I cut my marginal plants down in autumn?"

They will start to die back from now on anyway. Remove dead material regularly.

"What do I do with my water lily in autumn?"

Remove dead leaves as they appear. Green leaves should be left to die back naturally.

"Should I continue to feed my fish over winter?"

You will find the fish will get less active from autumn onwards, and will need little to eat. On warm days, they may come to the surface asking for food, and you can give them a small amount.

"Should I protect less hardy marginals over winter?"

The tops will start dying back as the weather gets colder, but the water will insulate less hardy marginal plants, like arum lilies and *Lobelia cardinalis*.

"Is there anything I should be doing with my wildlife pond at this time of year?"

Make sure that birds and small, non-hibernating mammals can reach the water for a drink. In a late, warm autumn, this may mean cutting back overgrown marginals early.

"My marginal plants are in crates and are becoming overcrowded. Can I split and replant them in the autumn?"

Growth slows as the water cools. This can cause root rot in newly planted aquatics, so it is best to wait until late spring.

"I have a lot of baby fish in my garden pond. Should I try and take them out now or wait until spring?"

If the bottom of your pond is not full of rotting vegetation, the fish will spend most of the winter down there, and a lot of the babies should survive.

"I have wet clay soil. Could I plant some trees in autumn?"

You might get problems with the roots rotting. Use the winter to improve the soil and plant in spring.

"My soil is heavy clay. Can I cover it in good topsoil before planting shrubs?"

No. You will get a layer between the different types of soil, which may drain badly and cause poor plant establishment. It is better to add sharp sand and organic material to your existing soil and plant when the texture has improved.

"How deep should I plant containerized shrubs?"

The general rule is to plant at the same level as they are in the container, with just enough soil covering the compost to prevent the area around the plant from drying out.

"What should I do with plants that are sent from online suppliers if I cannot deal with them straight away?"

Keep containerized plants outdoors and do not allow them to dry out. Bare-root plants will start arriving toward the end of autumn, and you can cover the roots with damp straw or sacking if you intend to plant within a day or two, otherwise plant them temporarily (heel them in) in a vacant piece of ground.

"Is it sufficient to dig a hole big enough to accommodate the rootball of a containerized plant?"

No. You should treat it in the same way as a bare-root one, digging a hole big enough to take the roots plus some good planting mixture. Firm in as you fill up the hole to prevent wind rock.

"I see many staking techniques in different gardens. Is one better than another?"

All gardeners have their own preferences. As long as the tree is firmly supported and does not rock in the wind, the technique is adequate.

"I heard that it is best not to stake a tree, as it encourages it to establish quicker if you do not support it. Is this true?"

This is not a good idea. The tree will develop a lean away from the prevailing wind, the head may break off, and, until a good root system forms, it will become loose in the soil every time the wind blows.

"How tall should a stake be?"

Ideally, you should be able to support the base of the head as well as the trunk. This is particularly important with top-grafted trees.

"If I use a vertical support to stake a containerized tree, where should I insert the point of the stake?"

You should position the stake as close to the trunk as possible without any part of it rubbing. This will mean knocking it through the rootball, but at this time of year, this should cause no lasting damage.

"Is one tree tie enough?"

It is best to use one near the top and one about halfway down. This will make sure that the trunk remains straight.

"When I'm planting, should I put manure in the base of the hole?"

Yes, but fork it into the soil at the base of the hole.

"Do I have to water the hole before planting?"

Unless your ground is moist, you should fill the hole with water and allow it to drain completely before planting.

"Is garden compost as good as farmyard manure when planting a shrub?"

Yes. Fork it into the base in the same way as manure.

"Do I have to water a containerized plant before I plant it?"

Yes. The rootball should be moist before planting.

"My bare-root roses have just arrived and the roots seem dry. What should I do?"

Soak them in a bucket of water for about two hours before planting.

"Should I water specimens immediately after planting at this time of year?"

If you are planting during a dry spell, you should give the area a good soaking after planting.

"Should I apply a mulch after planting in autumn?"

Mulching will keep warm soil at a higher temperature than normal for longer. As well as preventing moisture loss, it will help to establish the plants quicker.

"Is it a good idea to plant through landscape fabric?"

Geotextile will reduce the growth of weeds, especially in the short term.

"I think landscape fabric looks unsightly. How can I overcome this?"

Cover it with gravel, chippings or bark.

"I used landscape fabric last year, but weeds are starting to grow again. Should I lift the fabric?"

No. The weeds will be growing in soil that has collected on top of the fabric, but they will be much easier to remove.

"Do I plant first and then cover the ground with fabric or the other way round?"

You will find it much easier to cover the whole area with fabric first. When planting, cut a cross in the fabric so that you can turn it back to access the soil. Dig the hole (putting the soil on a piece of thick polythene, not the fabric), plant, then fill in the rest of the hole with some of the soil and replace the turned-back fabric around the base of the plant. Leftover soil can be spread thinly on beds not covered with fabric.

"What are cocoa shells used for?"

They can be used as a mulch or dug in as a soil improver before planting. As a mulch, the shells can become soggy during wet periods, so are best used around woody plants, like trees, shrubs, roses and fruit.

"I want to move a shrub that is in the wrong place. When is the best time to do this?"

Early autumn is best for evergreens; mid- to late autumn for deciduous shrubs.

"I am about to move a shrub that has been in my garden for five years. What special precautions should I take?"

Dig and prepare the hole at the new site first. Cut back tall top growth hard to prevent stress and wind rock. Dig the shrub out with as much soil as possible, then cut back any damaged roots, replant immediately and water well. You will need to water regularly during dry spells from spring to late autumn next year.

"Is it true you can get crocuses that flower in autumn?"

Crocus speciosus will flower from early autumn to midwinter according to cultivar. Plant corms in late summer in a sunny, well-drained spot.

"If I buy bulbs in late summer, can I plant them straight away?"

Yes. The only exceptions are tulips, which should be planted in late autumn or early winter.

"What should I avoid when I'm buying bulbs for spring?"

Avoid outlets where the bulbs are kept in warm conditions, because this will encourage premature growth and the spread of disease. Also, do not buy soft bulbs or those with long shoots already growing.

"Should I peel the brown skin from tulip bulbs before replanting them?"

No. The skin is there to protect the bulbs from damage and disease. Unless the skin is loose and comes off when handling, it should be left alone.

"I've seen spring bulbs on sale and it's only late summer. Surely this is too early?"

No. As long as the bulbs are dormant and not showing any signs of old foliage, this is the best time to buy them.

"I have bought several sacks of daffodil bulbs for naturalizing in grass. Should I plant each bulb individually with a bulb planter?"

Although this technique is recommended, it is much too much like hard work. Peel back the turf about 5 cm (2 in) thick so you can plant the bulbs in groups, lightly fork over the soil underneath, position the bulbs on the surface about 8–10 cm (3–4 in) apart and replace the turf carefully. Firm gently: any slight bumps will disappear as the bulbs start to grow.

"Is a bulb planter a useful piece of equipment at this time of year?"

No. It can be tiring if the soil is heavy, and it makes holes that are the same width and depth, regardless of whether you are planting a large bulb or a small one. Use a trowel instead.

"Some of my narcissus bulbs appear to be two bulbs joined together. Should I separate these before planting?"

You can separate them if they are the same size, but they will make a better show if left together. Large bulbs with small offsets should be planted intact; the offsets will not flower until they reach a certain size.

"I accidentally dug up some narcissus bulbs when I was clearing a border. They had started to shoot. I replanted them immediately. Was this the right thing to do?"

Yes. If this happens, try not to damage either the roots or the shoots, and do not be tempted to split the clumps by pulling them apart. Discard bulbs that have been cut in half.

"What is the difference between species and hybrid or Dutch crocuses? I've seen both types of corm for sale."

Species crocuses are those that either grow naturally in the wild or have been bred closely from them. They generally flower early. Large-flowered hybrids flower later and have bigger flowers. To get a show for as long as possible in spring, you need to have a selection of both.

"The crocuses in my mixed border are getting very overcrowded. When should I dig them up and split them?"

Autumn is the best time if you can find the dormant clumps, otherwise do it in early summer, just after the leaves have shrivelled.

"A friend has offered to give me some spring-flowering bulbs from her garden. Should I accept them?"

Find out if they are healthy and have been flowering well. If you do not know, it is better not to introduce them into your garden.

"I see on the labels of many bulb packets that most bulbs are offered for sale from cultivated stock. Why is this important?"

In some parts of the world, wild bulb stocks have been almost completely destroyed by being dug up and sold on for garden cultivation. In Britain, it is illegal to dig up wild bulbs, like bluebells, and replant or sell them on.

"Should I water my spring bulbs once I have planted them?"

This will not be necessary from autumn to spring, although watering after the flowers have faded will encourage better flowering the following year.

"I love the tall, large-flowered tulips, but in my garden they fall over, and this puts me off planting them. What can I do?"

Provide them with stakes at planting time, then you can tie them up before they start to flop. This will also help you to remember where you planted them.

"I have always planted lilies in spring, but an online site is offering them for sale now. Will they be all right if I plant them in autumn?"

As long as the soil or compost is free-draining and not likely to waterlog at any time during winter, they should be perfectly safe.

"Which bulbs can I plant in autumn for cut flowers next spring?"

Grape hyacinths (*Muscari*), narcissi, tulips, Dutch irises and star of Bethlehem (*Ornithogalum umbellatum*) bulbs are planted in autumn and make wonderful cut flowers.

"Last year, I planted some *Iris reticulata* in the rockery. Will they flower again next year or should I plant some more?"

After flowering, *Iris reticulata* bulbs often split into many small offsets. These may take a year or two to reach flowering size again, so if you want to see the flowers in your rock garden next year, you should plant some more.

"Is there any way I can make sure that dry snowdrop bulbs will flower next year?"

Some of them will take a year or two to come back into flower whatever you do, but if you buy and plant them as soon as they are offered for sale, they stand a better chance of flowering.

"Which is the small anemone I have seen flowering in woods in spring?"

This is the windflower (*Anemone nemorosa*). It can be grown in the garden in spring sun or light shade.

"Every autumn I plant florists' anemones, but they seldom come up the next year. What am I doing wrong?"

Anemone coronaria, with single cultivars in the De Caen Group and double cultivars in the St Brigid Group, need a warm, sheltered spot to thrive. They are often thought of as temporary and replanted every autumn or spring.

"When should I plant *Anemone coronaria* tubers?"

Plant them in autumn for flowering in spring, or in spring for flowering in summer and early autumn.

"How should I plant *Anemone coronaria* tubers?"

Plant them claws uppermost, 5 cm (2 in) deep and 8–10 cm (3–4 in) apart. Soak the tubers overnight to break dormancy.

"I plant tubers of winter aconites (*Eranthis*) every autumn, but never seem to get a good show. What am I doing wrong?"

These are often better grown in pots and planted out when they are in full flower and leaf in spring. They seed readily after flowering, so never hoe round them once the flowers have faded.

"How can I grow some dog's tooth violets?"

Erythronium are not easy. The corms should be planted in autumn, or they can be potted up and planted as young plants in spring. They will not thrive unless the soil is humus-rich, moisture-retentive and neutral or slightly acid.

"Can I plant oxalis in my rockery this autumn or should I buy plants next spring?"

First, be sure of which oxalis you want to grow, as many species are invasive and almost impossible to remove. Plant *Oxalis enneaphylla* as dormant corms now, in a sunny, well-drained part of the rockery.

"My flat has a tiny, sunny balcony. Can you suggest some plants for winter interest?"

Winter heathers and dwarf conifers will give a good display and last for many years, and winter-flowering pansies bloom from autumn until early summer.

"Can I leave my new cordyline outdoors over winter?"

For new cordyline, it would be better to take it into a greenhouse, cool conservatory or unheated room, just for this winter.

"What bulbs should I be planting in my containers for spring flowering?"

Crocuses, dwarf daffodils and botanical tulips will give you 'a bright splash' from late winter and throughout the spring.

"Should I take my 60 cm (24 in) high cabbage palm (*Cordyline australis*) into the greenhouse for the winter?"

No. Unless it is exceptionally cold, it will survive outdoors.

"Can I plant bulbs in hanging baskets for the spring?"

Crocuses, dwarf narcissi and shorter tulips do well in winter hanging baskets.

"Last year I planted daffodils in hanging baskets and they looked untidy. What can I do with the bulbs?"

Taller narcissi tend to fall over in hanging baskets, so replant the bulbs in a large container or tub.

"Is it possible to have a semi-permanently planted winter hanging basket?"

Yes. Use the largest solid-sided basket you can find, and fill it with soil-less compost. Suitable plants are winter heathers, primulas and polyanthus, dwarf thymes and other evergreen herbs, and periwinkle. Do not overplant, and add new pansies and forget-me-nots every autumn. A basket like this should last about three years.

"I want to plant some bulbs in containers. How can I get the best possible show?"

Plant in layers, starting with the largest bulbs (standard narcissi and tulips) first, and ending with the small bulbs, like snowdrops (*Galanthus*). This will give you flowers in a single container from midwinter to late spring longer if you add some alliums.

"Will bulbs in containers flower the second year?"

If you use a soil-based potting compost and keep them well fed and watered until the foliage dies, they will come up and flower for several years.

"Can I plant pansies on top of spring bulbs in a container?"

It depends on the bulbs. Crocuses will work well, but bulbs with large leaves will smother the pansies.

"When I'm replacing summer bedding with winter and spring plants, can I use the same compost?"

You will find that the top layer will be full of roots and much of it will come up with the discarded bedding, so it is better to start again. If you have deep pots, however, replacing the top half with new compost is usually adequate.

"Last winter, the compost in my outdoor pots got soggy and many plants died. Should I add something to the soil to prevent this?"

No. Improve the drainage in autumn by checking that the holes in the container bases are not blocked. Use pot feet or bricks to raise the containers off the ground.

"I have some wire hanging baskets. Should I plant the sides as well when I'm planting them up in autumn?"

Wire hanging baskets are not the best for autumn planting because they freeze more quickly and the plants in the sides often get damaged by strong winds. Invest in some solid-sided ones for overwintering outdoors, or keep the baskets in a cold greenhouse until early spring.

"I have just bought a camellia for a container. Do I have to keep it indoors until the spring?"

As long as the nursery kept it outdoors, you can keep it outdoors too: just make sure it is in a sheltered spot.

"Are terracotta pots really frost-proof?"

Some are more frost-resistant than others, but no ceramic pot is as durable as wood or best-quality plastic.

"I have a new patio and want to get some containers on it as soon as possible. What should I look for at this time of year?"

Fibreglass, heavy-duty plastic and wood containers have good insulation properties and are less likely to be affected by frost than ceramic ones. As with all things, you get what you pay for.

"Will an olive tree survive outdoors if I buy it now?"

Olives are hardier than you would imagine. A standard olive should survive quite happily outdoors in winter, but small ones need protection for a year or two.

"I bought a variegated lemon tree this summer. The label said it was hardy, so can I leave it outside in winter?"

Citrus trees are usually hardy to only a couple of degrees of frost. If you live in an area that is likely to get spells colder than this, it would be wise to give it winter protection.

"I have an alpine sink garden. Is there anything I should do with it now?"

Make sure it will get as much sun as possible during the shortest months of the year. Check the drainage holes and apply a mulch of grit or pea shingle to keep the leaves off damp compost.

"Should I feed my containers in autumn?"

This should not be necessary because active growth is slowing down rapidly now.

"Should I use a container compost when planting up in autumn?"

Many proprietary container composts contain water-retaining substances that may make the compost too wet in winter. Use multi-purpose compost instead.

"Can you suggest indestructible plants for a winter hanging basket on a cold wall?"

Use variegated, small-leaved ivies (*Hedera*) in soil-based compost. They will withstand these conditions well and can be clipped back in spring for permanent use if required.

"I have a lot of hostas in pots on the patio. Now the leaves have died, the pots look bare. What should I do with them?"

These will survive quite happily in an inconspicuous place, such as behind a garage or shed, from now until the spring. Replace them temporarily with pots of winter and spring bedding, and bring them back when growth starts to show in spring.

"When I plant up my baskets in autumn, do I need to water them with a vine weevil control?"

This will also control other early pests, like pansy aphid, so it is a good idea.

"My garden store has cyclamen for sale for planting in containers now. Will they survive outdoors?"

If they have been properly hardened off, they will be quite happy in cold weather, but more than a degree or so of frost will kill the flowers.

"I have a window box containing a 'New Wave' trailing petunia that survived outdoors all last winter and has been magnificent this summer. Is this unusual?"

Although we treat them as annuals, petunias are really perennials. The window box will be getting some heat through the window, and the house wall is warmer than the open garden. Out of interest, see if it survives another winter.

"I have seen small wallflower (*Erysimum*) plants for sale in packs of six. Would these be better for my tubs than the bare-root plants?"

Not usually. These are immature plants that are unlikely to flower well next spring and need another summer's growth to make good flowering plants.

"What plants other than wallflowers can I plant in my spring tubs?"

Try Brompton stocks (*Matthiola incana*) or, for a follow-on display, sweet Williams (*Dianthus barbatus*).

"I have a large, overgrown garden. What is the easiest way to keep it under control? I am a pensioner and do not have a lot of money."

You may have to spend some money to start with, but it will save time and effort in the long run. Get a landscape gardener to clear your plot of everything you do not want. Make large beds of shrubs and cover the rest with grass. You can control weeds among shrubs with a thick layer of bark mulch, and you will find that modern lawn mowers are light and easy to handle.

"The bottom of my garden is boggy. What can I plant there?"

An easy and inexpensive approach would be to have a bog garden. Plants like candelabra primulas, marsh marigold (*Caltha*), rodgersia, *Lobelia cardinalis* and arum lilies would love these conditions, and provide interest from spring to autumn.

"How can I quickly cover a large fence?"

Try *Lonicera japonica* 'Halliana', a fast-growing, fragrant honeysuckle that can be cut back fairly hard once it has covered the fence.

"What flowering plants can I grow in shade under conifers?"

No flowering plants grow well in this situation. Instead, use tubs of bedding plants and plant some spare ones to replace the originals when necessary. They will soon revive in a sunny spot and can be reused.

"Can you suggest an ornamental tree for a small garden, 6 x 11 m (20 x 35 ft)? I would like to have a sitting area beneath it."

Try a standard form of the small crab apple, *Malus floribunda*. The spreading branches will shade the sitting area from the sun.

"I have a number of dogs, but would like to improve my garden. Can you suggest some dog-proof shrubs for my borders?"

The urine of dogs will burn evergreen foliage and kill herbaceous plants, so choose deciduous shrubs like forsythia, lilac (*Syringa*), deutzia, *Viburnum opulus* 'Roseum' (sometimes called 'Sterile'), mock orange (*Philadelphus*) and shrub roses, and keep the dogs away from them while they establish. If your dogs like to chew, prickly shrubs like berberis and holly will soon put them off.

"Are there any other types of ceanothus than the blue-flowered cultivars?"

Ceanothus × pallidus 'Marie Simon' has pink flowers and C. × pallidus 'Perle Rose' has rose- carmine ones. Both are deciduous, flowering all summer with reddish-brown bark in winter. C. thyrsiflorus 'Snow Flurry' has white flowers and glossy, evergreen foliage.

"Can you suggest a shrub for a cold, windy wall?"

Plant the dwarf shrub *Euonymus* × *fortunei* 'Variegatus' at the base of the wall. In this position, it will start to climb with its suckers.

"I need a windbreak, but the area is subjected to salt-laden winds. What would survive here?"

Any of the ornamental hawthorns, such as *Crataegus laevigata*, will survive the salt. You could also try any cultivars of mountain ash (Sorbus aucuparia) or whitebeam (S. aria).

"What non-poisonous plants can I include in my new garden? I have two small children."

Most plants can cause problems in some ways, either a tummy upset, an allergy or a scratch. Tell your children not to eat or touch anything unless you say so. They will soon learn to keep away from prickles and thorns.

"What can I plant in an area of damp shade?"

Why not have a fernery? Most common ferns love these conditions, and the effect can be quite stunning.

"I don't seem to have much luck with magnolias. Is there an easy one?"

Magnolia 'Susan' will tolerate most soils. It makes a small tree quite quickly and flowers from an early age.

"Can you recommend some vandal-proof shrubs?"

Rubus cockburnianus, Berberis julianae, Rosa rugosa and sea buckthorn (Hippophae rhamnoides) should do the trick.

"Can you suggest a hedging plant to keep next door's dog out of my garden?"

Hawthorn (*Crataegus monogyna*) or blackthorn (*Prunus spinosa*), planted in a double row with 45 cm (18 in) between plants, will eventually do the trick, but you will need to put up a temporary wire-netting fence while the hedge thickens.

"Is there a conifer that will grow in damp soil?"

Swamp cypress (*Taxodium distichum*) will tolerate constantly soggy conditions. It is not evergreen, but the foliage turns bronze in autumn and the bark is an attractive shade of red.

"Is autumn the best time to make a rockery?"

It is a good time to do the construction work, but unless you plant up early in the season, some alpines may struggle over winter. Construct the rock garden now, let it settle over winter and add plants in the spring.

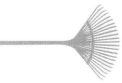

"Will I damage a wooden-boarded fence if I grow climbers up it?"

Self-clinging climbers, like Virginia creeper (*Parthenocissus*), *Hydrangea anomala* subsp. *petiolaris* and ivy (*Hedera*), will make it difficult for you to get at the boards to apply preservative. Other climbers can be supported by a removable trellis, which can be detached and pulled forwards so that you can get to the fence if necessary.

"Which small tree could I plant as a feature in a courtyard covered with gravel?"

The willow-leaved or weeping pear (*Pyrus salicifolia* 'Pendula') has silver leaves and white flowers in spring. It looks lovely in a courtyard, where the branches can touch the ground without spoiling other plants or grass.

"I have a small front garden that is open to the road. How can I define the frontage without boxing it in with a fence?"

Why not use railings, either on a low wall or down to ground level? These will not block the view, but will give the garden some protection from the road.

"How can I make a straight path to my front door look more interesting?"

Make a feature of it by flanking it with a dwarf hedge of lavender, interspersed at intervals with standard roses all of the same cultivar.

"How can I make my long, narrow back garden more interesting?"

Use trellises, rose arches or shrubs to divide it up into a series of 'rooms', but leave broad walkways through to attract the eye. Now is the time to start planning.

"I read recently that you should plan your garden on paper. Why?"

It is easy to plant too closely and to forget to allow room for growth. When you are positioning the plants on paper, it is easier to allow for their sizes in, say, five or ten years to avoid overcrowding problems.

"Which ornamental shrubs would look at home in a wildlife garden?"

Try the wayfaring tree (*Viburnum lantana*) and guelder rose (*V. opulus*). Hazel (*Corylus avellana*) and golden elder (*Sambucus nigra* 'Aurea') would also be appropriate in a such a garden.

"I have chalky soil. Can I grow a rhododendron?"

You could remove the earth from an area and replace it with acid soil. Alternatively, you could grow a rhododendron in a large container. However, your mains water is likely to be alkaline, so you must make provision for collecting enough rainwater to see it through dry spells.

"I would like a *Buddleja davidii* to attract butterflies. Don't these shrubs get too large for small gardens?"

The cultivars 'Nanho Blue' and 'Nanho Purple' seldom grow more than 1.2 m (4 ft) tall if they are pruned hard each spring.

"How do I plant a living hedge?"

Take long, unrooted shoots of willow (*Salix*) or red dogwood (*Cornus*) and push them firmly into the ground in autumn. Most will root. Next summer, weave them into each other to create an impenetrable screen.

"I have a small lawn. Should I be thinking of something else in this area?"

When a lawn gets to be so small that it serves no useful purpose, it is often better to take it up and cover the area with ornamental chippings. Surrounding plants can then spread out without ruining the lawn edges.

"What can I use to remove green algae and lichen from my fruit trees?"

The chemicals that were used for this until a few years ago have been taken off the market in many countries. In fact, the algae and lichen on fruit trees will do no harm, and are only an indication that the air is clean and healthy.

"Is autumn a good time to plant strawberries?"

The best time to plant bare-root strawberry plants is between late summer and mid-autumn. Container-grown plants can be put in at any time when the weather is suitable.

"What shall I do about the masses of runners on my strawberry plants?"

These can be removed in autumn, if you have not already done so, to keep the parent plants individual. However, if the runners are left to root into the bed, you will get a much heavier crop next season, although many of the fruits will be smaller.

"When should I cut back fruited strawberry plants?"

Late summer or early autumn is the time to cut back the plants and generally tidy up the bed.

"Can I propagate from my raspberry plants?"

If the plants are young and healthy, you can replant suckers in late autumn.

"What's the best way to make some new blackberry plants?"

By layering. Bury the tip of a healthy shoot of this season's growth in early autumn and peg it down to keep it firm. The tip will root and may be severed from the parent plant and planted in early autumn next year.

"How can I take blackcurrant cuttings?"

Take cuttings, 25 cm (10 in) long, from healthy, heavy-bearing bushes in autumn. Bury them so only two buds show above the soil.

"I have some good gooseberry bushes I raised from cuttings last year. What pruning do they need now?"

Remove the lower branches so that the bush grows on a short 'leg', and cut back the side-shoots you retain by about half. Remove the leading shoot completely.

"I have been given a small plant of a Worcesterberry (*Divaricatum* hybrid), which looks like a cross between a gooseberry and a blackcurrant. How do I look after it?"

This is actually a species of currant, but is thorny and is treated like a gooseberry.

"When can I expect my melons to ripen?"

They usually ripen from early autumn onwards, depending on variety.

"How do I grow a kiwi fruit?"

These plants, *Actinidia deliciosa*, are deciduous climbers, and they need quite a lot of space. You will either need both a female and a male plant for fertilization, or you can choose a self-fertile cultivar, like 'Jenny'. Plant against a sheltered, sunny wall or fence.

"When should I plant a container-grown, fan-trained peach tree?"

From early to mid-autumn is the best time, but you can plant later in the year if conditions are mild.

"How much space should I allow between a fan-trained peach and a fan-trained nectarine?"

Leave about 4 m (12 ft) between trees to allow for growth.

"What are mirabelle plums used for?"

These sweet mini-plums can be used in the same way as ordinary plums or made into wine.

"Why do my apple trees produce a lot of top growth every year?"

You are probably winter-pruning too hard. Leave them unpruned this year and see if growth slows down.

"I have limited room to grow fruit trees and would like to plant some cordons. What is the minimum space between plants?"

Allow 75 cm (30 in) between trees.

"I love apples but, only have room for one tree. Are there any self-fertile varieties?"

If you have good soil, try 'Self-Fertile Cox'. Alternatively, plant 'Redsleeves', which is partly self-fertile.

"Which is the best culinary crab apple?"

Malus 'John Downie' has attractive flowers and crops heavily. The fruit makes good jelly.

"Are the fruits on my 'Maypole' ballerina apple tree edible?"

Yes. They will make a pinkish-red jelly.

"My local nursery offers 'step-over' apples. What are these?"

These are effectively single-tier espaliers trained horizontally about 30 cm (12 in) above the ground. They are pruned like espaliers in late summer or early autumn, and make an attractive edging to a bed or border.

"What are water shoots?"

These shoots are vigorous, soft and unproductive in the short term. They are often caused by over-severe pruning, and the majority should be removed in autumn, although one or two well-placed ones may be allowed to remain and will eventually fruit.

"When I prune apple and pear trees, how can I identify a fruit spur?"

Fruit spurs are short, twiggy branches bearing fat buds that will bear fruit. They produce no extension growth, so will get no longer.

"When are the fruits of a medlar ripe?"

Pick the fruits in late autumn and leave them for many weeks until they turn brown. This process is known as bletting.

"Can I grow new potatoes for midwinter?"

If you have an unheated greenhouse, cold frame or conservatory, you can have new potatoes on Christmas Day (in the northern hemisphere). In late summer or early autumn, buy seed potatoes that have been specially treated (most big seed companies offer these). Put two or three on a layer of multi-purpose compost in a large pot and fill the rest of the pot with compost. Keep damp but do not overwater, and shoots will soon appear.

"When should I lift maincrop potatoes for storage?"

The best time is early to mid-autumn, but if the weather is wet, you may have to lift later.

"When should I sow overwintering broad (fava) beans and early peas?"

Mid- to late autumn is the best time. Cover them with fleece to protect them from birds.

"When I pulled up my runner beans, the roots looked rather like dahlia tubers. Is this a disease?"

No. Runner beans are, in fact, perennial plants, and if the tuberous roots are protected from frost, they will come up again the following year, although the yield will not be as good.

"An experienced gardener told me I should not remove pea and bean roots at the end of the season. Why?"

These should be left in the ground as a useful source of nitrogen.

"I sowed some perpetual spinach and still have a lot of leaves left. It seems a shame to pull them up. What should I do with them?"

Perpetual spinach is winter hardy. If you cover the crop with fleece, you should be able to pick sparingly throughout winter, and you will get some early pickings next spring, although the plants will eventually run to seed.

"Do I have to lift beetroot (beet) in autumn?"

If you protect the crop with fleece or cloches it may survive the winter, but it is likely to be damaged by slugs. If you do not have the space to store the crop, however, it is worth the risk.

"Will my main-crop carrots be all right if I leave them in the ground until I need them?"

They will survive the winter, but so will overwintering pests, such as aphids and carrot fly grubs, so it is better to lift them.

"Can I use cloves from my home-grown garlic to grow another crop next year?"

Yes, as long as they are absolutely disease free.

"When should I sow Japanese onions?"

Sow seed in early autumn or plant sets in late autumn.

"I love turnip tops. Can I sow now for an early crop next spring?"

Sow a maincrop variety in early autumn for spring greens from early spring onwards.

"When should I plant rhubarb?"

Traditionally, early spring was the time for planting, but on well-drained soil and in warmer districts, you can plant from late autumn throughout the winter.

"What should I do with globe artichoke plants at the end of the season?"

Remove any unharvested heads and stems and any dead leaves at the base of the plants. Do not plant offsets until spring, or they may rot.

"What is the latest time I can sow winter radishes?"

Late summer is the best time, but you can still sow in early autumn if you have had to wait for space in the vegetable garden.

"I have grown some sprouting broccoli plants. When should I plant them in their final positions?"

Plant whenever they are ready from early to late autumn. Protect them against birds.

"How far apart should I space spring cabbage?"

Space plants about 10 cm (4 in) apart and thin out gradually in spring. Use the first thinnings as spring greens.

"Why does spring cabbage never form good hearts?"

This is due to wind rock. Earth up the plants in autumn to prevent this.

"I grew celery this year for the first time, and I found that many of the plants were rotten when I harvested them. What went wrong?"

This is a disease called celery heart rot. It is caused by a bacterium that enters the stem through a wound, and there is no cure. Cultivate around the plants carefully, and keep slugs under control. Do not compost diseased plants and do not grow celery on land where the disease has been a problem the previous year.

"Can I sow lettuce in autumn for an early crop next year?"

Choose a winter-hardy variety, such as 'All the Year Round' or 'Winter Density'. In colder areas, the plants will need protection with cloches.

"I have grown far too many winter squashes. Will these store?"

You can keep them in a cool, dry place for several months.

"My parsnips have got brownish-black areas near the top. What caused this?"

This is parsnip canker. It is worse in early-sown crops and acid soil. Lime the area before growing parsnips again, and choose a canker-resistant variety. Do not compost diseased roots.

"Why did all my cauliflowers develop brown curds this season?"

This is usually a sign of boron deficiency. Add plenty of garden compost in autumn, and apply borax before planting next year.

"My autumn cabbages are covered in whitefly. What can I do?"

Spray with a vegetable insecticide at three-day intervals until the problem is resolved. Do not compost infested leaves.

"Why is my kohl rabi tough and stringy?"

You are leaving it too long before harvesting. The globes should be no more than the size of a tennis ball, and preferably golf-ball size, to be at their best.

"What is the best way to store kohl rabi for the winter?"

Kohl rabi does not store well and should be eaten fresh. Leave plants in the ground until they are needed.

"How long should I expect my tomato plants to last?"

They will last well into the autumn if you look after them properly. Pick the fruit regularly.

"When is the best time to clean out the greenhouse?"

The gap between clearing out vegetables, such as cucumbers and tomatoes, and bringing in your overwintering patio plants is the time for an autumn clean. Wash off any shading and pay particular attention to moss growing between the glass and the glazing bars. Use a greenhouse disinfectant to clean the glass, staging and floor.

"Many of my tomatoes are struggling to ripen as the leaves are shading the fruit. What should I do?"

Regularly remove and clear up all leaves that are covering the trusses.

"What can I do with the green tomatoes left at the end of the season?"

Ripen them indoors on a sunny windowsill, or put them in a plastic bag with a ripe banana. Remove old plants, growing bags and spent compost immediately and wash the pots (if used) before storing for the winter.

"When should I be thinking about insulating the greenhouse?"

The best time is immediately after it has been cleaned out.

"My greenhouse is full of permanent plants. Can I clean it out without removing them?"

Only if you use a disinfectant that states there is no need to remove the plants first. Even so, it will be more difficult to clean every part thoroughly.

"I grew a pineapple plant from the top of a fruit this summer. Can I overwinter it in an unheated greenhouse?"

No. It will probably be too cold. Put it on a sunny windowsill in the house in autumn.

"I have just bought a gardenia. Where should I keep it over winter?"

A heated conservatory or light windowsill indoors is best.

"Many of my conservatory plants need repotting. Is it too late to do this?"

Growth will be slowing up now, so putting the plants in larger pots of new compost can cause the roots to struggle. Wait until spring.

"We often hear that you should bring houseplants inside the room when drawing the curtains at night. Is this still true with double glazing?"

This will depend on the plants in question, the thickness of the gap between the two panes of glass (the smaller the gap, the less effective the insulation) and whether you have net curtains, which can save up to 25 per cent of the heat lost through the glass. If you are in doubt, move them.

"Are there any bedding plants I can sow in autumn in the greenhouse?"

You can sow pansies and sweet peas (Lathyrus odoratus) at this time of year.

"Can I sow any vegetables in autumn in an unheated greenhouse?"

Carrots, lettuce, radishes and land cress can be sown now for an early crop.

"When can I take pelargonium cuttings?"

The second half of summer is best, but you can continue to take cuttings until mid-autumn if you do not overwater them.

"Why did most of my pelargonium cuttings rot last winter?"

Leave the cuttings for a day to dry out at the bottom before inserting them into compost. Add some sharp sand to multi-purpose compost when you are potting up, and take care not to overwater.

"Can I take shrub cuttings in the greenhouse at this time of year?"

Autumn is the best time for taking cuttings of evergreen shrubs, but they are quite difficult to do, so take more than you need. It may be some months before you know if you have been successful.

"When I put my spring bedding plants in, I am left with a lot of perfectly good summer bedding plants, such as *Begonia semperflorens* and coleus (*Solenostemon*). What I can do with these, apart from consigning them to the compost heap?"

Many summer bedding plants can be potted up at the end of the season and taken into the conservatory or greenhouse, or even put on a light windowsill indoors. They will then continue to give pleasure for several more weeks.

"When is the best time to bring in greenhouse and conservatory plants that have spent the summer outside?"

Do not be caught by an early frost. As soon as you have room for them in late summer or early autumn move them inside.

"Why do I seem to get botrytis in my greenhouse in autumn?"

Water only when necessary and do the job carefully. Do not wet the staging, floor or leaves once the weather starts to cool down.

"How can I stop my greenhouse getting steamy on sunny days in late autumn?"

Don't be afraid to ventilate. Use the roof ventilators and any windows, but do not leave the door open. Close up again in late afternoon.

"What easy plants can I buy to brighten up my conservatory at this time of the year? I cannot afford much heat."

Most garden stores and some DIY stores stock ornamental peppers, winter cherry (*Solanum capsicastrum*), indoor chrysanthemums and florists' cyclamen (*C. persicum*). If you can maintain a temperature of no less than 5°C (40°F), they should be quite happy.

"Can you suggest some shrubby plants for my cool conservatory from now until spring?"

Hebes, variegated cultivars of *Euonymus japonica*, *Pittosporum* and *Griselinia* can be successfully grown under glass during this period, as long as the temperature does not get too high.

"Can I plant a grape vine this autumn as shade for my sunny greenhouse?"

Yes, but ideally the vine should have its roots outdoors, so you may have to make a hole somewhere near the base. It should be trained to shade the sunny side only.

"Is it possible to grow a Christmas-tree-like conifer in my conservatory that I can decorate at Christmas?"

Look for the Norfolk Island pine (*Araucaria heterophylla*). Although a relative of the monkey puzzle tree (*A. araucana*), it resembles a young Norway spruce (*Picea abies*) and can be decorated as such. Keep it in a cool, light spot from now until spring.

"Should I think about bringing my bonsai plants in until the spring?"

It depends on the species. Many, like Japanese maples (*Acer*), conifers and zelkova, are hardy and should be kept cold and allowed to become dormant in winter. Others, like bougainvillea, *Ficus benjamina* and pomegranate, cannot tolerate frost and should be given a fairly warm, light position indoors from now on.

"I would like to have some lily-of-the-valley (*Convallaria majalis*) in my conservatory for the spring. How can I make sure they flower at the right time?"

Pot up some crowns from the garden in mid-autumn. They will flower in the conservatory in late winter.

"My camellia is getting too large and heavy to move into my greenhouse in autumn. Can I leave it outdoors?"

As long as it is in a position where the flowers will not get early sun in spring, it will be quite safe outdoors.

winter

"I can never get a poinsettia to survive for long when I buy one in winter. All the leaves drop off within a matter of days. What am I doing wrong?"

Give the best light possible and never let the compost dry out, but do not leave the pot standing in water. It will then last for several months.

"Are poinsettias with white or pink flowers harder to keep growing?"

Not really, providing they are given the right conditions.

"What do I do with an indoor azalea (Rhododendron simsii) when the flowers have faded?"

Remove spent flowerheads and continue to water, feeding fortnightly. Repot in ericaceous (acidic) compost in spring.

"Why have the leaves of my indoor azalea turned yellow?"

You need to use soft (rain) water for watering. Give it an ericaceous plant tonic and it should recover.

"I forgot to water my indoor azalea and the leaves have shrivelled. Will it survive?"

No. It has probably died. Never allow these plants to dry out, even for a short time.

"How do I get the bracts on a poinsettia to turn red in time for Christmas?"

Eight weeks before you want the bracts to turn red cover, the plant in a black bin bag or similar at night to give 14 hours a day of total darkness. After eight weeks, the plant should be treated as normal.

"Why do none of my indoor cyclamen last more than a week or two?"

Florists' cyclamen (*Cyclamen persicum*) need a cool, light position. They will suffer in a hot room.

"Do I have to water my cyclamen from the bottom?"

This is often suggested as the only method, but as long as you keep the water off the corm, it is not necessary.

"Should I repot my amaryllis (*Hippeastrum*) after flowering?"

No. It will flower better in following years if it is potbound.

"What do I do with the dead flower spikes on my amaryllis?"

Cut them off completely, near the base.

"What should I do with my amaryllis bulb after flowering?"

Continue to feed and water it until late summer, then stop watering. The leaves will die and can be cut off. Store the dry bulb in a cool, dry place until late winter.

"Can I plant a winter cherry (*Solanum capsicastrum*) outdoors?"

This indoor winter plant used to be discarded when the berries dropped off, but with warmer winters, it will usually survive in a warm, sheltered position outside.

"I've been given a planted bowl. What do I do when the plant finishes flowering?"

Take the plants out in spring and repot them individually.

"Can I prune an over-large Christmas tree (*Picea abies*) in my garden?"

You can prune lightly, but it will alter the shape.

"Will my rooted Christmas tree survive if I plant it out?"

If it has a good root system and is not losing its leaves, it will probably survive.

"If I get my Christmas tree to grow in the garden, can I reuse it indoors every year?"

No. Doing this will put too much strain on an already over-stressed plant.

"How can I put my large, unrooted Christmas tree to good use after the holiday?"

Fix it in the ground with a metal post spike and use it as a bird-feeding station.

"What can I give a fussy gardener?"

How about a subscription to a plant society or for a gardening magazine?

"When should I cut back my hellebores?"

Wait until they have seeded, because you will get many seedlings that can be used elsewhere in the garden, then remove all old leaves and flower stalks.

"Why do my Christmas roses (*Helleborus niger*) always flower at different times?"

Some strains come into flower earlier than others, and much depends on the temperature. Most will flower at some time from late winter through spring, and covering them with a cloche will encourage them to flower earlier.

"How can I stop my pulmonarias looking so lack-lustre after flowering?"

Remove the old flower stalks after seeding and cut off the old leaves. The plants will soon produce new, well-marked foliage.

"What can I buy a keen wildlife gardener?"

A bird feeder, a supply of bird food, a lacewing chamber, or toad hibernation shelter would all be acceptable.

"Can I grow mistletoe (*Viscum album*) in the garden?"

Try pushing some berries into the bark on the younger branches of apple, hawthorn, lime or poplar trees in spring.

"There is a big mistletoe plant on my apple tree. Why does it never produce berries?"

It is almost certainly a male plant and never will.

"Can I grow mistletoe from cuttings?"

No. Mistletoe is a parasite and must be grown from seed on a suitable host.

"I have a large holly tree in the garden. Why does it never produce berries?"

You have a male plant. Buy a female, such as *Ilex aquifolium* 'Madame Briot'.

"How can I stop birds stripping all the berries off my holly?"

In future, decide in autumn which branches you want to cut and net them first.

"Can I move a seedling holly?"

Older hollies (*Ilex*) do not move well, but young ones can be moved successfully in late winter or early spring.

"Is there a tree that will be in flower for a Northern-hemisphere Christmas?"

The winter-flowering cherry (*Prunus* × *subhirtella* 'Autumnalis') should be in flower unless the winter is very cold.

"How and when can I prune back an overgrown white rose trained over an arch?"

In winter, remove all dead, dying and weak wood. Cut back some older shoots hard in order to encourage new growth, and trim the shoots that have borne flowers to about 10 cm (4 in) from the base.

"What is the correct way to remove a big branch from a tree?"

Make an undercut first to prevent the bark from peeling back as the branch starts to fall. Make the cut slightly proud of the trunk or main branch to help the wound callous over.

"Should I paint the cuts when I remove branches from my trees?"

No. They will heal quicker without.

"How do I prune a silver birch so as not to spoil its shape?"

Betula pendula does not shape up well after its head is pruned, so restrict pruning to raising the crown by removing the lower branches.

"Can I use shredded prunings on the garden straight away as a mulch?"

Yes, but only if there are no signs of disease on them.

"I pruned my silver birch hard last winter and it never grew again."

Silver birch should be pruned only when it is at its most dormant, just after the leaves have fallen. At other times, it is likely to bleed to death through excess sap loss.

"The trees in our garden are now getting too big and blocking the light. Can I lop them back?"

Yes, but you will probably spoil the shape. Also, the roots could damage the foundations and drains, so check that there are no restrictions on their removal before having them taken out and replaced with something more suitable

"My young silver birch has branches almost to the ground. Can I remove some of these so that I can see the bark better?"

Yes. Do the job in early winter.

"What is pollarding?"

This is the process of cutting back the head of a tree hard. It is often done to get the best bark effect in trees with attractive bark, but can also be done to prolong the life of an old tree, especially willows (*Salix*).

"I took some branches off my flowering cherry in early winter, and now I learn I should have done the job in summer. Have I killed the tree?"

Possibly not. Paint the cuts with a fungicidal pruning compound and keep your fingers crossed.

"My *Robinia pseudoacacia* 'Frisia' flowers well, but every winter a lot of the branches break in the wind. Should I replace it?"

You may be better growing it for its foliage alone. Cut the head back hard this winter, then trim the new branches back every year.

"Can I cut back my leggy penstemons?"

Wait until you see new growth coming from low down on the stems in late winter or early spring, then cut back to 5–8 cm (2–3 in) from the base.

"When should I prune an overgrown *Mahonia japonica*?"

Prune immediately after flowering, which is generally in late winter.

"How should I winter-prune an overgrown apple tree?"

Remove all but about five, evenly spaced main branches. Then thin out the branches coming from these so there is good air circulation in the head and no shoots are crossing or rubbing against each other.

"Which shrubs should I winter-prune?"

There are few shrubs that should be regularly pruned in winter, but overgrown deciduous ones can be cut back hard and shaped at this time.

"My outdoor ornamental vine (*Vitis vinifera* 'Purpurea') is threatening to engulf the house. Can I cut it back hard?"

Yes. Do the job in the first half of winter, before the sap starts to rise.

"How should I prune a winter jasmine (*Jasminum nudiflorum*)?"

After flowering, prune back all the flowered shoots to the main branch framework.

"I have an ornamental vine (*Vitis coignetiae*). How should I prune it?"

Prune all side-shoots back in winter, leaving one or two dormant leaf buds on each shoot. In summer, shorten all new growth to one or two leaves beyond the fruits that are forming.

"Is it necessary to spend a vast amount of money on a pair of secateurs?"

A few years ago, the answer would have been yes, but these days there are some good, inexpensive secateurs around. Make sure the blade is always sharp and positioned correctly.

"What is a node?"

A node is a dormant growth bud on a stem or branch.

"What are the rights and wrongs of pruning dormant trees and shrubs?"

Always remember to prune to a node, from which new shoots will be produced. A long piece of wood above the node will not shoot and is likely to die back, often into healthy tissue.

"When I'm pruning roses and similar plants, should the cut slope toward or away from the bud?"

The cut should slope downwards, away from the bud, so that water will drain off and not rot the bud.

"Is it possible to remove the bottom branches of a mature *Chamaecyparis lawsoniana* 'Alumii' so that I can plant underneath it?"

Yes. You can do this at this time of year, but the shape may not be as attractive. You may also find it difficult to get anything established in the dry, shady soil around a conifer like this.

"I have a large clump of bamboo, which collects dead leaves from the nearby trees. I want to clean it up. Can I cut it all back and start again?"

Yes. It will usually resprout in spring.

"Is it true that you shouldn't remove the old heads of hydrangea bushes until after the risk of frost has passed?"

This used to be recommended because the dead heads protected the new, flowering shoots underneath from frost. As we now seem to be experiencing warmer winters though, it is unlikely the new growth will be damaged if you want to tidy the plants earlier.

"When should I clip back the ivy (*Hedera*) on my house wall?"

Wait until late winter, when the birds will have eaten all the berries and they are not in need of a sheltered place to roost.

"Why does my soil develop a green slime on it in winter?"

It may be 'panned' (trodden hard) or it could be the type of soil. Clay and silt soils settle down with rain to a smooth, hard surface. Dig in winter to improve the texture.

"Should I remove ivy from my trees and hedge?"

Out-of-control ivy can eventually smother a hedge and the younger branches of trees and shrubs. In deciduous hedges, it can be managed by spraying with glyphosate while the shoots are bare. Otherwise, cut it off at the base and treat new growth with glyphosate as it appears. If you prefer, allow it just to grow a little way up the trunk and trim away excess growth regularly.

"Should I be adding any fertilizers in winter?"

The only concentrated fertilizer likely to be useful is bone meal, but winter is a good time to dig in bulky manures and compost.

"What exactly is soil?"

Soil is made up of four basic ingredients: mineral particles, organic material, air and water. Good soil should contain all of these in the right proportions.

"I am told I have a silt soil. What is this?"

Silt soils have small particles, but not as small as those of clay. They are usually fertile and reasonably easy to work in good weather.

"Why should I not use nitrogen-rich fertilizers in winter?"

Nitrogen promotes growth, and during the dormant season most plants will not be able to take it up. It washes out easily and can become a water pollutant.

"I have a heavy clay soil. What should I be aware of when planting trees and shrubs?"

If you use a planting mixture for back-filling that is more free-draining than the soil, the soil water will drain into this like a sump and the roots may rot. Instead, add a small amount of sharp sand to your garden soil to encourage roots to become established.

"Should I put lime on the soil every winter? Can you add too much lime?"

Unless you have peaty or extremely acid soil, liming every three years is adequate.

"Why do clematis not grow well in my peaty soil?"

Clematis and many other ornamental plants need an alkaline soil to thrive. Add a handful of lime when planting and top-dress with lime in spring.

"Why is humus in the soil a good thing?"

Humus contains beneficial soil organisms. It also holds moisture and allows air into the soil.

"Why should I not use peat as a soil conditioner?"

Excess peat makes the soil acid, light and quick to dry out in summer. Remember, too, that extracting peat does not help the environment from which it was taken.

"Is there any way I can prevent my compost bin from cooling off in winter?"

Wrap the outside in several layers of greenhouse bubble insulation and cover the lid with an old carpet or similar.

"I put my garden compost on the roses every winter. Why do they never look particularly healthy?"

You need a large amount of compost or manure to give most plants all the nutrients they need and supplementary feeding during the growing season with blood, fish and bone, a balanced fertilizer or a specific plant fertilizer is nearly always necessary.

"How much manure or compost should I add in winter?"

A bucketful per square metre (yard) is sufficient. This will provide a small amount of nitrates, phosphates and potash, and some trace elements.

"I want to make a heather bed. What materials should I add to my soil?"

Because the use of peat is now discouraged on environmental grounds, the best substitute is leaf mould (mold).

"I always apply a mulch of leaf mould (mold) in winter. Why are my garden pinks dying off?"

It will tend to make the soil acid and pinks (*Dianthus*) prefer alkaline conditions. Test the soil and lime it to give a pH of over 7 before replanting.

"What is a runner bean trench?"

This is a trench at least 30 cm (12 in) deep, which is dug in autumn. Shredded newspaper, kitchen waste (other than cooked food, meat and dairy products) and other organic material can be added throughout the trench. When full, it is covered with soil. As it rots down, it both warms the soil and improves the texture.

"I've seen mushroom compost for sale. Is it good for the soil?"

Because it contains some lime, as well as organic material and compost, this is a good mulch for all except lime-hating (ericaceous) plants.

"I keep chickens. What is the best way to usethe manure from the poultry house?"

Add it to the compost heap or bin as an accelerator.

"A farmer has given me some old straw. Should I dig it in?"

Compost it first, in case it contains agricultural weedkillers.

"What is gypsum?"

Gypsum (calcium sulphate) is an organic fertilizer containing sulphur and calcium. It can be used to lighten heavy clay soil because it causes the fine particles to stick together.

"What material other than peat can I use to make my soil less alkaline?"

Sulphur is a good alternative. Test your soil and follow the instructions on the packet.

"Should soil be turned over during digging, or is it enough just to break it up?"

If it is turned over the birds, will eat pests, like cutworms and chafer grubs. It will also improve aeration.

"What are forced and non-forced bulbs?"

Forced bulbs are given a cold, dark period after planting to get a good root system growing. Non-forced bulbs are treated in the same way as if they were planted in the garden, and will flower later in the season.

"What are prepared hyacinths?"

These are bulbs that have been temperature treated for early flowering indoors.

"What should I plant indoor bulbs in?"

A soil-less type of compost that has had charcoal added to keep it sweet. This is generally used when the pot has no drainage holes.

"How do I plant up tulips for indoor bowls?"

Cover the bulbs with compost to only half their depth.

"How can I grow indoor narcissi?"

Plant in late summer or early autumn in bowls. Put them in a cold, dark, frost-free spot (such as in a black bin bag in the garage) and bring them into a cool room when the shoots are 2.5–5 cm (1–2 in) high.

"How close should I plant tulip bulbs in a pot for indoor flowering?"

The bulbs should be close together but not touching.

"Is it all right to use general-purpose compost for indoor flowering bulbs?"

Yes, but you must make sure the pots have adequate drainage holes at the base.

"What spring bulbs can be potted up to flower indoors?"

All spring bulbs can be used for indoor decoration as long as they are properly treated after potting.

"I notice that the tips of potted-up bulbs for sale in garden stores are uncovered. Why is this?"

You should leave the tips uncovered if you intend to force the bulbs for early flowering. Otherwise, they should be covered.

"I have no suitably dark place indoors to keep pots of bulbs after planting. What can I do?"

As long as you do not want the flowers early, you can pot the bulbs up in multi-purpose compost in containers with drainage holes and leave them outside until the flower buds are showing.

"Can I grow a hyacinth in water?"

Use a specially designed hyacinth glass. Fill the bottom section with water and put the bulb in the top section, with the base just touching the water. Place in a black bag in a cold place and bring it out when it shows about 2.5 cm (1 in) of growth.

"Can bulbs other than hyacinths be grown in water?"

Non-hardy forms of narcissus, like those in the Tazetta group, can be grown in pebbles in a glass bowl containing just enough water to keep the bulb bases wet. There is no need to cover them to exclude light.

"What should I do with bulbs that have flowered indoors?"

After flowering, move them to a cool, light place and feed and water until the leaves start to yellow. Then plant them in the garden. You can either do this straight away or dry them off for autumn planting.

"Do I plant bulbs that have flowered indoors at the same depth as in the pots?"

No. They should be planted at the depth they would have been if they had been planted outdoors in the first place: that is, three times as deep as the height of the bulb.

"I planted a lot of the bulbs that had flowered indoors outside in the garden. Why did they not flower the following spring?"

Growing them indoors will have given them a shock that they need time to recover from. Feed the leaves with a bulb or balanced fertilizer and they will flower again in years to come.

"Are there any indoor flowering bulbs I cannot plant outside afterwards?"

Tender narcissi, like *N. papyraceus* ('Paper White') and 'Grand Soleil d'Or', would not survive the frost. Hyacinth bulbs grown in water should be discarded after flowering. 'Prepared' hyacinths may never produce flowers as big outside.

"I have bought some hyacinth bulbs for indoors, but I do not think they were 'prepared'."

Don't worry. They will flower just as well, although later in the season.

"Can I use bulbs that have flowered indoors for a second year inside the house?"

No. They will not give a good display.

"How can I stop indoor bulbs flopping over when they start to flower?"

Insert short pieces of split cane and tie the stems to these. Alternatively, ring all the growth round with soft string or wool.

"I planted blue, pink and white hyacinths in a bowl. Why did they all flower at different times?"

They probably would. The only way you can get them to flower at the same time is to pot each bulb up individually and transfer those with flowers at the same stage to the bowl just before flowering.

"Is it possible to mix different kinds of bulbs, such as daffodils and tulips, in the same bowl?"

No. One or the other kind will flower first, and its foliage will spoil the effect of the later flowering bulbs.

"Why do my indoor narcissi have really long, pale, floppy leaves?"

This is what happens when they are placed in a warm room in a spot without full light.

"Why are my hyacinths flowering at the bottom of the leaves?"

The most likely cause is that they were brought indoors too early.

"Why are my forced hyacinths short and stunted?"

Either you did not water the compost enough before they were brought indoors, or they were not kept in the dark for long enough.

"Why do the buds on my indoor hyacinths often fail to open?"

Erratic watering is usually the cause, or you might have wet the developing buds when you watered.

"What causes deformed hyacinth flowers?"

This happens when the pots have been kept too warm after planting, or when they have not been given a period of darkness.

"Should I knock the snow off the roof of my greenhouse?"

Yes. Snow will cut down light drastically and put undue strain on the framework and glass.

"What is the best heater for a small greenhouse?"

A greenhouse fan heater distributes the heat most evenly and can also be used to circulate unheated air in summer.

"What ornamental plants can I grow in my unheated greenhouse in winter?"

Hardy annuals in pots will give you early spring interest. Many ferns can be displayed under glass in winter, then moved outdoors in summer. Indoor cyclamen will flower if the greenhouse is insulated

"How warm should my greenhouse be in winter?"

If it is heated to a minimum of 7°C (45°F), you can grow a wide range of frost-free plants and overwinter patio specimens. Fuel costs usually prohibit a higher temperature.

"I have a large greenhouse and cannot afford to heat it. What can I do?"

Consider dividing off a small area with heavy-duty polythene and heating just that. You could use electric heating cables or a heated propagator to start off seeds early.

"Which shrubs will flower in winter in an unheated greenhouse?"

Try camellias and indoor azaleas. Move them outside in summer.

"Will cacti survive in an unheated greenhouse?"

If they freeze, they will die. However, they will withstand temperatures down to freezing if they are not watered. Move them away from the glass.

"What can I use to insulate my greenhouse?"

Bubble insulation (with large bubbles) gives the best protection. Buying it as packing material from an office stationer is much cheaper than getting it from a garden store.

"The inside of my greenhouse is covered in green algae. Should I remove it?"

If you use your greenhouse in winter, you must remove this regularly because it will severely reduce the already poor natural light.

"Should I water the plants in my greenhouse in winter?"

Only water when the compost becomes dry. Try not to wet the floor or any gravel on which the plants are standing. Water in the morning in mild weather.

"I want to do something different with my unheated greenhouse. What do you suggest?"

An alpine house would allow you to grow all the alpines that do not like cold, wet winters. The only essential requirement is good ventilation at all times.

"The older leaves of my overwintering patio plants are covered with mould (mold). What should I do?"

Always clear away dead and dying leaves and do not touch unaffected leaves until you have washed your hands. Open the ventilators on sunny days.

"Are there any vegetables I can sow now?"

In a frost-free greenhouse, you can sow early cabbage, lettuce, summer cauliflowers and leeks for an early crop.

"Can you suggest some scented plants for my conservatory?"

Hyacinths, *Narcissus papyraceus* ('Paper White') and indoor jasmine (*Jasminum polyanthum*) will fill your conservatory with fragrance. Citrus plants produce sweetly scented flowers throughout the year.

"What do I do with a large indoor jasmine (*Jasminum polyanthum*) that is flowering at the moment?"

Cut it back hard in early spring. Wind the new shoots round a hoop or trellis as they grow.

"I have a heated conservatory. Can I sow tomato seed for an early crop?"

If you can maintain a minimum night temperature of 10°C (50°F), you can sow the seed in a propagator for cropping from early summer onwards.

"Why are the leaves of the bougainvillea growing in my cool conservatory falling off?"

If the temperature drops below 5°C (40°F), it will become deciduous (lose its leaves in winter). Reduce watering to a minimum and prune back hard in early spring.

"When should I start cutting back overwintering fuchsias, *Plectranthus* and pelargoniums in my cool greenhouse?"

As soon as you see new leaves, you can prune these plants back hard and repot them.

"I keep some fuchsias, pelargoniums and other plants in my greenhouse all year, but after repotting I am running out of space. What should I do?"

When you are repotting, cut the roots back until they will fit back into the old pot with a little space for new compost.

"How can I prevent a container-grown peach on my patio from getting leaf curl every year?"

Move it into a cold greenhouse or conservatory for the winter to keep the rain off it because the spores that cause it are carried in rainwater.

"I am sure I saw silver snail trails in my conservatory the other day. Is this likely?"

When the greenhouse warms up in winter, sunshine slugs and snails will become active. Put a few pellets where you saw the trails.

"What jobs should I be doing now in preparation for spring in the greenhouse?"

Wash all empty containers, pots and trays with a garden disinfectant. Make seed labels out of old plastic bottles and order seeds for sowing in spring.

"Can I get an early crop of strawberries in the greenhouse?"

Pot up some plants from the garden and put them in a light position on the bench or staging. You will get a considerably earlier crop.

"What can I use my cold frame for in winter?"

Sow salad leaves in autumn, and use the frame to protect them. 'Amsterdam Forcing' carrots will germinate in growing bags under a cold frame in late winter.

"When can I take cuttings of my tender fuchsias?"

Take these as soon as shoots appear that are long enough to use as cuttings, usually from late winter on.

"How do I take chrysanthemum cuttings in the greenhouse?"

In midwinter, take cuttings about 8 cm (3 in) long of basal growths (not side-shoots).

"Why did my seed-grown pelargoniums not flower until late summer last year?"

You probably sowed rather late. Sow in a heated propagator in winter and prick out when large enough, and at a temperature of no less than 13–15°C (55–60°F).

"Are pots of ready-to-prick-out seedlings a good buy?"

Yes, but only if you can be sure they have been looked after properly by the supplier, and you have a heated greenhouse at about 15–17°C (60–65°F).

"I have a tall cabbage palm (*Cordyline australis*), which catches a lot of cold winds. Is it too big to cover up? Will it survive?"

The older it is, the hardier it becomes, so it is unlikely to suffer. Strong winds can actually be beneficial because they will strip out all the old, dead foliage and spent flower stalks.

"Should I tie up the leaves of my cordyline over winter?"

In cold areas, this will protect the central leaves. Tie them up loosely and untie in early spring.

"Last winter, my tubs froze solid and many plants died."

Wrap the containers in two layers of bubble insulation.

"Should I cover my less hardy plants with polythene over winter?"

No, they might rot. Use a double sheet of horticultural fleece instead.

"Why do many of the alpines in my rock garden die in winter?"

Cover them with open-ended cloches or a piece of glass to stop the rain falling directly on them.

"How do I keep my tender fuchsias over winter without a greenhouse?"

Dig a trench in the garden and bury them. Remember to mark the trench so that you know where they are.

"How can I keep a standard fuchsia alive in my unheated greenhouse?"

Wrap the vulnerable stem in pipe lagging over winter.

"Is it possible to keep pelargoniums from year to year without a greenhouse?"

Shake off some soil or compost and pack them into boxes. Keep them in a frost-free place, such as a garage.

"Is it true you can save pelargoniums by hanging them upside down?"

If you hang them up in a frost-free place, with a little soil or compost still on the roots, you may save up to 90 per cent of them.

"Should I cut my half-hardy patio plants back before I take them into the greenhouse for winter?"

No. They may rot back from the cuts. Wait until growth starts in spring.

"How can I stop my dahlia tubers from rotting ?"

Remove the rotten ones immediately and check the rest for soundness. Dahlia tubers should always be stored upside down so that the remaining sap can drain out. Dust with fungicide.

"Is it necessary to take up half-hardy plants, such as dahlias and gladioli, for the winter?"

In warmer areas, half-hardy plants might overwinter successfully in the ground if you cover them with a mulch of straw or compost. Remove this in spring.

"I have been storing my main-crop potatoes in old compost bags. Why have they all gone rotten?"

Potatoes must be stored in opaque, breathable linen sacks. Some seed companies offer these for sale.

" Can I have half-hardy patio plants outdoors in winter without ill effect?"

Tender fuchsias, pelargoniums, marguerites and osteospermums may be quite happy outdoors in winter in mild areas. However, if you value them, it is best to give them some protection.

"How can I keep my osteospermums alive without a greenhouse?"

The newer cultivars are almost hardy, but if the temperature falls sharply, you should cover them with fleece or old net curtains.

"What should I do with my pond pump in winter?"

Many manufacturers recommend that you leave the pump in the pond permanently. Run it from time to time to make sure it will still work.

"I have a wildlife pond without fish. Should I keep it free of ice?"

If it is deep enough, it is not necessary to remove ice.

"I planted a new herbaceous border in the autumn. Does it need winter protection?"

For the first winter after planting, cover it lightly with straw or fleece. This will not be necessary in subsequent winters.

"I planted a new conifer hedge last year, but it has turned brown on the windward side. How can I protect it?"

Plastic windbreak material is available, or you could protect it from the wind with hurdles or bamboo panels. You can remove these when the plants are established.

"I pruned my roses in autumn and they now have shoots on them. What should I do to prevent frost damage?"

Rose shoots produced in colder weather are much hardier than those appearing after spring pruning. It is unlikely, therefore, that frost will do any serious damage.

"I have a peach tree trained against a wall. I am told to keep rain off it in winter to prevent peach leaf curl. What is the best way to do this?"

Erect a temporary cover of heavy-duty plastic sheeting over the top so the rain runs off it to the sides and front. Remove the cover in spring.

"What is the best way to look after garden furniture in winter?"

Garden furniture is best stored indoors when it's not in use, but if this is not possible, you will find a range of waterproof covers for tables, chairs, hammocks, barbecues and the like in large garden centres and in DIY stores. Remove these regularly to check that the furniture is all right – mice, rats and other creatures appreciate the protection, too.

"Do I have to mow the lawn in winter?"

They can be topped if the weather is mild and dry.

"How can I get a supply of mint throughout the winter?"

Pot up a few roots and put them in a warm, light place, such as the conservatory or on the kitchen windowsill. New shoots will soon appear.

"Do my containers need watering in winter?"

Containers in the open usually get watered naturally by rain in winter, but in a long dry spell you should check the compost to see if it is dry. Containers in the shelter of walls and under eaves may need watering occasionally.

"My garden ornaments turn green in winter. How can I clean them?"

A pressure washer or patio cleaner will do the trick. If you are using chemicals, keep them off your plants. Some moss or algae on ornaments can look attractive, but keep birdbaths and bird tables clean.

"Is there ever a time in winter when there is nothing to do in the garden?"

No, not really. There are tree stakes and ties to be checked, fences to be mended, paths to repair and clean, hanging basket brackets to make safe...as well as planning for next summer. Happy gardening!

Index

Index

Index

Index

Index